Exodus

Book Two of
When We Were Gods

Daniel Seltzer

Two Cents Plain Publishing Co.

Exodus is a work of fiction. Names, characters, places, and incidents are the products of the author's imagination or are used fictitiously. Any resemblance to actual events, locales, or persons, living or dead, is entirely coincidental.

2025

ISBN 978-0-9898045-2-3

eBook ISBN 978-0-9898045-3-0

Printed in the United States of America

Cover Design by

CMS Graphic Design

EXODUS

Dedicated to
My sister, Karen Mesi, and my friend, Jim Walter, both of whom kept on me to continue work on the series.
And, of course, to those of you who are patient and kind enough to take the time to read this book.

Some things are more precious because they don't last long.

Oscar Wilde - *The Picture of Dorian Gray*

CHAPTER ONE

Niklas Mueller.

It was the second time in the past ten minutes that Clay heard his iMeme whisper the name into his ear. Each time he felt a wave of fear sweep over his body

Think, damnit! Think!

The last several hours had swept past Clay. So much so, they seemed to have never even come, let alone gone.

Where am I? he asked. Silence.

Jesus, Clay. Get a grip, he admonished himself. His now shut-down iMeme was unable to respond to his inquiry. *Am I that reliant on my stupid iMeme that I can't get by without it?* He had always disliked the iMeme, a convenience he could live without. But now, it seemed to have become a necessity with which he could not live without, but which had suddenly become too dangerous to live with. He looked at the watch on his wrist. The display showed it was 2:17 a.m.

He did not recognize the watch. It was not his. He tried to think. *Maybe Eva gave it to me before I left her apartment? When was that? Late Thursday evening, early Friday morning?* Clay tried to recall. *Was this one of Miguel's old watches? When did she give it to him.* His memory could not retrieve the moment. The watch was too big for Clay's wrist. He doubted that Eva ever had worn it.

Jesus, what am I doing here? He remembered most of Thursday evening vividly. Eva. The failed meeting with the Walking Man. The Niklases. Lt. Cobb. But after that, it was sketchy.

What Clay did know was that while he had intended to go to his inoculation appointment, he somehow never made it. Even now, he still held deep reservations. But there was no way to avoid it without deep repercussions. This he felt as surely as the watch on his wrist.

* * *

The moment he stepped out of his house on Friday morning, Clay had the sense of being watched. He looked around, but the streets were empty. But still, he could not shake the feeling. He quickly turned his head and thought he caught the glimmer of a man stepping into a shadow. *Was that a Niklas?* He paused to change direction to see. *No. But what could I do if it were?* The hair on the back of his neck stood on end. This scenario happened twice again, and each time a sense of dread arose. Clay approached the corner he did not recognize. The street signs read WOOD and CONGRESS PARKWAY. Rush medical center was just ahead. *Is this where I'm supposed to be? Dr. Kovak's office isn't here at Rush?* Rather than continuing straight to the hospital, he turned south. *If someone is following me,* he thought, *it should be easier to figure out if I take a round-about route.* He continued for a block south and turned back east. The road dead-ended into what appeared to be the back of a concrete and red brick building. *Isn't this Harrison? I should be able to take this all the way into the Loop, but – ?* Clay turned back around but again the street seemed unfamiliar. The expressway noises rose in the distance and he started walking toward the sound, trying to regain his bearings. At Congress he started walking east again. He walked for several blocks but never passed the hospital. He looked ahead at the street sign on the next corner. RACINE. *What happened? I had to have passed by the hospital.* The sky was clear and Clay felt the heat of the sun on the back of neck. An uneasy feeling swept over him. "What time is it?" Clay asked.

It is six o'clock p.m., his iMeme responded.

"Six p.m.! What the – ?" He turned to look behind. He spotted two Niklases standing just inside a nearby doorway. He turned his head quickly and tried to walk as calmly as he could northbound on Racine. *Did they see me? Do they know I saw them?* The fear rose inside and he broke into a trot to the Blue Line subway entrance and sprinted down the ramp to the platform. He reached the platform, out of breath and an ache forming in his side just as a train was prepared to disembark. Clay jumped aboard. As the doors closed and the train lurched forward, he saw a group of three or four Niklases coming down the subway ramp. He crouched down quickly, garnering stares from the several persons seated in the railcar. *Was it his age? or his actions?* He did not know. Clay waited until the train cleared the station and entered the dark tunnel below the street before he got up and started moving between cars towards the back of the train. The last car was empty. *Thank God!* Clay breathed a quick sigh of relief. He walked to the rear door. As the train entered a yellow slow zone, it started to decelerate. *Now's my chance!* he opened the rear door and jumped.

He landed hard on his right side and quickly rolled onto his stomach. He knew that the third-rail lay to his right, and he had no desire to electrocute himself. The pain in his right shoulder was intense, but he sensed he had little time before more Nicklases would be boarding the train at the next stop, looking for him. He got up to his feet in the dark and edged toward the inner wall of the tunnel. He knew that the walls contained regular set-backs designed to allow workers a safe place to stand clear of trains while working the tunnels. While these set-backs would offer protection from being struck by a train, he knew that it would not take long for the Nicklases to figure out that he was in the tunnel system. Clay shuffled deeper into the tunnel, looking for one of the several emergency exits leading out of the subway and up to the street. A dimly lit sign notified him that he had found what he was looking for and he managed to return to street level without being accosted by Niklases. But he needed to find shelter.

Clay understood that he would not easily blend into a crowd. It was easy enough to spot him due to his age; the fact that he had just jumped off the back of a moving 'L' train did not help. The only saving grace was that it was already dusk, and the emergency exit seemed to have let out into an alley. Although Clay could not place his location, he had an uneasy feeling that safety lie at home. *But I am so tired, and the pain*. He looked around. The alley was empty, except for several dumpsters and a few discarded pallets.

Maybe I can just sit here for a moment. Catch my breath. Then I can figure out what to do, he thought. He sat down between the two dumpsters and closed his eyes.

When Clay reopened his eyes it was dark and quiet. Too quiet for the city. He could discern no noise except for the low murmur of someone talking and several sets of boots against the pavement. Two men walked past the alley.

Who is that? he thought.

Niklas Mueller, his iMeme responded.

"Shit!" he whispered a little too loudly for comfort. He grabbed his iMeme and shut it down. *If I can I.D. them, they can I.D. me.* Clay waited until he thought the Niklases had moved past and he quickly stood up and ducked into an alley doorway. His arm was throbbing as he checked the back door of the building. It gave in to his slight tug.

That's odd, he thought. But he did not wait to question it. *My luck is rolling good today.* He slowly opened the door and then eased it back closed so that the door would shut gently and quietly. He was in some type of storage room filled with shelves and boxes. There was a layer of dust on everything and it appeared that the business that last operated there had left some time ago.

He made his way across the darkened room to a door and slowly opened it a crack. As he did so, he heard footsteps from further down the hallway and quickly, but gently, closed the door. He pulled his iMeme from his pocket, turned it on and slipped it under the door, just at the front edge and waited.

Who's coming? he repeated to himself as he heard the footsteps pass.

Niklas Mueller, his iMeme responded.

He slipped his fingers under the door to retrieve his iMeme and he quickly shut it down.

Think, Damnit! Think! A moment passed and he wondered, *Where am I?* Silence.

Jesus, Clay. Get a grip, he admonished himself. His now shut-down iMeme was unable to respond to his inquiry. *Am I that reliant on my stupid iMeme that I can't get by without it?* He looked at the watch on his wrist. The display showed it was 2:17 a.m. He did not recognize the watch.

Jesus, what am I doing here?

Clay tried to remember when he had left Eva's place. *Late Thursday evening, early Friday morning?*

He got up slowly and walked back across the room and carefully exited to the alley. *Trapped,* he thought. He rubbed his shoulder, which felt better and noticed that his clothes did not seem as ragged as he had earlier imagined them. He looked down the alley and was surprised to see the sidewalks were filled with people. Perhaps if he could not blend into the crowd, he could at least count on sheer numbers to shield him. He stepped out of the alley and turned left. As he began walking he noticed that everyone else seemed to be headed in the opposite direction. He began to struggle like a swimmer fighting a rip tide. He considered turning around but he needed to get back home and while he wasn't sure where he was, he somehow knew he had to keep moving in this direction. He bent his head down to avoid eye-contact. A couple of times he thought he heard his name, but he continued on. When his fear grew too much, he slipped his hand into his pocket and turned on his iMeme. He slipped his hand out of his pocket, thinking *Who is it?* as he gathered the courage to raise his eyes for a moment.

Clay slammed head-on into another pedestrian and he lost his grip on his iMeme. As it slipped out of his grip, Clay watched it falling seemingly in slow motion, turning over and over, and it whispered to him before it shattered on the sidewalk at his feet.

Niklas Mueller.

Clay found himself sitting in a chair facing a large video screen. Although he was not restrained in any way, he did not attempt to rise, nor even turn to see the faces of the men standing beside his chair. He knew who they were. Niklas Mueller and Col. Leeds.

"So you see, Mr. Furstman, it is not necessary that you actually tell us what you may know about . . . what do you call him?" Leeds paused. "Oh yes, the Walking Man." There was another pause. "We can access that information directly from your brain. We can capture your thoughts. Read your mind, so to speak." Leeds paused again. "It can be an entirely painless process. But of course, we wouldn't want you to miss out on any of the fun, would we?" Leeds did not wait for a response. "No. We have quite a few surprises in store for you."

Clay still did not turn towards Leeds, but he could feel a smile form across Leeds' face.

"Yes, quite a few painful surprises, Clay." He moved away from Clay's chair and towards the video screen. "Do you mind if I call you Clay? I feel that since we will soon be sharing your deepest secrets that we should consider ourselves . . . not quite friends, but at least on a first-name basis. Is that okay with you, Clay?"

Clay remained silent.

"I know you are concerned about your family. I want to assure you that they are being taken care of." At this, the video screen jumped to life and Clay saw four people scaling a steep cliff.

"Did you know your son and one of your daughters won a trip to Colorado to learn to rock climb?" At this, the camera zoomed in on the four climbers, slowly scaling the rocks dozens of feet above a river. Two climbers were acting as lead climber, one to each of his children, who were slowly ascending from below. Both lead climbers had already anchored cams just above a ledge and were watching Clay's children scale the rock face. Due to the architecture of the cliff face, Clay's children were attempting to scale an overhang and both children were relying entirely on their ropes to pull themselves up to the ledge where the lead climbers sat.

"Your almost there, keep going," Clay could hear one of the lead climbers shout.

"Grip the ropes tight," the other added.

Matthew looked up and smiled toward the men, while Lizzie seemed to be struggling a bit to pull her body up the rope. The two men then looked at each other and, almost simultaneously pulled knives from their belts. Clay saw that they were each a Niklas. They smiled at one another and nodded. Each swiftly slashed the rope stretched out below them. Clay watched in horror as the taut ropes suddenly went lax. Matthew and Lizzie, plunging toward the river below, were screaming. Each had a length of rope from their harnesses trailing above. Clay was struck with the bizarre notion of a world where cutting the umbilical cord, rather than giving life, caused death

"You do know, Clay, don't you, that we have the ability to shut all those little nanobots down? The ability to return pain and disfigurement to the human body? The same as would have been done to Finkelstein and Landsburg had they been found guilty." Leeds continued to stare at the video screen. "Even to suffer death."

Clay opened his mouth to speak, but Leeds interrupted him. "No. There was nothing you could have done. This is not a live video feed, Clay. What you see here has already come to pass. You see, we don't need your cooperation, Clay. Merely your physical presence."

A numbness passed through Clay's body.

"You're probably wondering about your wife and other daughter, Katie?" Leeds mused.

For the first time, Clay turned towards Leeds. He was filled with anger and loss but seemed unable to rise.

"You need not worry, they are not dead," Leeds continued. The scene changed on the video screen. A creature, humanoid in shape, lay upon a bed, its arms held near its ribcage, fists clenched. Clay looked upon it in wonder. The neck was too long to be human. Clay guessed it was at least eleven inches and it looked far too thin to support the hairless head, which was larger than that of a human; macrocephalic. In the center of the neck was a large red welt. The creature appeared to be trying to pull its head into its torso, to shrink the length of its neck, while a nurse kept stroking the neck, stretching it out to its full length. All the while, the nurse was conversing with another woman, her back to Clay, who was nodding, apparently familiar with what was taking place. The women spoke too softly for Clay to hear what was being said, but Clay got the impression that if the creature's neck was allowed to contract to its shoulders, the creature would die. Clay suddenly recognized that the woman with her back to him was Lillian. A chill ran through Clay's body as he looked back towards the creature. He fought hard to swallow the vomit that had leaped into his throat. As he scanned the creature's face he was drawn to its eyes. They were unmistakable.

Katie! Clay wanted to scream, but he was unable. He was transfixed on her eyes. Haunting. Filled with anger and sadness at the cruelty which she clearly recognized her life had become. Eyes that held disbelief, pain, and, perhaps, hate. Eyes that screamed for answers. *What's going on? Must I live like this forever? Must I live at all? Why me?*

How many times have I told her that life wasn't fair? Clay wondered. But he never meant anything like this. *What would become of her life? What could?* As Katie lay there, struggling to pull her head down, fists clenched in anger or struggle or pain, the nurse and Lillian stroking her neck back to full length, Clay sensed in Katie a desire to end the suffering. *Is she struggling to contract her neck by reflex? pain? Does she know it is a way to end the pain? to die?* Clay briefly tried to comprehend the effort and wondered. He tried to drive a thought from his head, but couldn't help himself: *Do I hope she finds success*?

How can you wish death on your own child? He chided himself. Lillian's resolute was apparent. It was clear she did not want to lose Katie and was willing to undertake what was necessary for their daughter. Clay sensed the love Lillian kept in her heart, the love of a mother to a child, grateful that her child was alive, nodding in understanding of what needed to be done. Lillian seemed to have accepted the cruel fate delivered and, Clay knew, unable to even consider other options. Clay continued to stare into his daughters eyes, unable to turn away.

"You see, Clay, PreVentall can be used not only to prevent illness, to correct DNA errors, but to change DNA; to create a new genome, change to who we are, what we are to become," Leeds said, his voice revealing no sympathy. Clay moved to leap from his chair, but Niklas' hands were already upon him, holding him down.

"And now, Clay, it's your turn," Leeds said, holding a large stainless steel device. Clay felt the grip of a large forceful hand cupping his skull from behind and pushing his chin towards his chest. Clay screamed as he struggled against the pressure, unwilling, or unable, to remove his gaze from that of Katie's suffering eyes.

* * *

Clay sat bolt upright in his bed, his body covered in sweat. He looked over at his clock. It was 5:23 a.m. Clay let out a sigh of relief. *It was only a dream!* He laid back and closed his eyes, but quickly opened them again. He was not looking forward to being inoculated later this day. In fact, he was overcome with a feeling that he was making a terrible mistake. But it was not the PreVentall that was keeping him awake. He tried closing his eyes one more time, but his mind was haunted by the image of Katie's eyes.

CHAPTER TWO

RING. . . . RING. . . . RING. . . . RING. . . . RING.

After the fifth ring, the phone fell back silent. The caller would have been transferred to voicemail and given the option to leave a message. Eva felt safe to open her eyes and look at the clock. It was 4:18 a.m. A clear glass bottle was sitting on its side, the neck hanging over the side of the table. It was upended as a result of an unsuccessful attempt to set it down several hours earlier. She had passed out on the couch, her robe draped over her half-naked body like a blanket. She felt drained. The couple of hours she had slept were insufficient to have provided any relief from the exhaustion she had felt come over her body since yesterday evening. The steadiness of the digital numbers shining at her, however, told her that the short rest was long enough to free herself from the tilt-a-world atmosphere that she had been recklessly tumbling toward. Her head, tightening its grip on her consciousness, begged for hydration. She was just beginning to consider getting up to appease her head for the sake of more sleep when the phone began to ring again.

"Shit," Eva muttered as she rose slowly to answer the phone. She knew before she answered whose voice lie at the other end of the receiver and she had no desire to answer it, but knew that she had little choice. A slight chill settled over her and she bent down to pick the robe up from where it had fallen and pulled it tightly around her. She felt an odd comfort in knowing that her robe had been wrapped around Clay only a few hours earlier. As if by inhabiting the intimate space of her robe, Clay had left a part of himself which offered her closeness and protection. Eva picked up the receiver.

"Hello," she said. Her voice rough from troubled sleep. And alcohol.

"Professor Diaz, I trust all is well with you?" There was no attempt at apology or explanation for the hour of the call, nor for response to the question. "It seems that a situation has developed and I'm afraid that it is necessary for you to report immediately."

"Col. Leeds," Eva began, the adrenaline now flowing in her bloodstream providing a level of mental clarity that did not exist only moments earlier. The events of last evening flashed through her mind and she wondered what Lt. Cobb may have reported, or the Niklases, and whether Leeds was toying her her, like a cat before killing its prey. "Why, thank you for asking. I am doing well, although I must admit that I was sleeping before you called."

"Forgive me, my dear Eva. I guess it is very early in the morning back in the States isn't it? But, you see, I am caught up in some very serious business and you will excuse my indiscretion. It is imperative that you join me. I will bring you up to speed then."

Eva let out a silent sigh of relief. Leeds' call apparently had nothing to do with the events of last night. "Of course, Colonel. I don't mean to be curt with you. I apologize, I am still half-asleep, but can't this wait until morning?"

"Morning has past. It's after noon where I am and I need you here. I have sent a car to pick you up and drive you to the base. You can be ready in twenty minutes, I trust." Leeds quietly commanded.

Eva looked around the room. She was not prepared to pick up and leave at a moment's notice. "Twenty minutes? It's going to take me that long to get dressed, let alone pack and get ready to leave for . . . " Eva paused. "Just where is it I'm going and for how long?"

"That, my dear, is classified information at this time. All your needs will be provided here, making your departure quite simple."

"It's not that easy Colonel. I . . . I . . . ." Thoughts began to race through her mind. She was not ready, nor perhaps willing, to pick up and leave to some unknown destination, not knowing what may have become of Professor Nakosh, of her run-in with Cobbs, of Clay. "I need more time, Colonel," she said, less sure of herself than usual.

Her demand was greeted with silence. She wrapped the robe more tightly around her body as a chill passed through her. She took another breath, prepared to renew her request when Leeds spoke.

"Well, I certainly don't want to upset my favorite professor. I will send a car over to pick you up oh-six hundred thirty hours. You will be ready by then." With that Leeds hung up.

Eva let out a long, loud sigh as she set the phone down. She walked into her bathroom, and immediately began to dress. Her head begged for a little more sleep before she set out, but she reminded herself that she could rest during the trip to wherever Leeds had determined she was headed.

She drafted a short note and placed it, along with a key to her home, in an envelope and sealed it shut. She wrote CLAY hurriedly across the front on an odd slant in order to avoid writing over the key. She folded the envelope and placed it in her pocket as she stepped into the hallway, quietly closing the door to her unit behind her. In the lobby of her building she paused, glancing out into the pre-dawn light to see if anyone was lurking about. The street was empty and she slipped out into the night.

Eva awoke with a start as her alarm chimed. It was almost 6 a.m. She had been dreaming and she awoke with a feeling that she had something important to do. Something in her dreams that needed to be completed. She remembered that a Niklas or two would be coming to pick her up in less than an hour, but the dream seemed to beckon something more. She sat in bed a moment, trying to sort out her dream, but as she reached out to grab the thought, it seemed to break apart, leaving her holding only a fragment of memory, a disassociated collection of remembrances that refused to take on a coherent shape. She remembered struggling, but over what she could not recall. A feeling that something was missing swept over her body. She looked around her room, considered what she needed to do, but everything seemed to be in order

The sunlight filtered into her room and she got up to take a quick shower, the empty feeling still nagging at her. It was as if in forgetting her dream she was letting someone down, failing in an important task that needed to be completed. She washed her hair and then massaged in conditioner, letting it sit in her hair while she lathered up the rest of her body. She tried to shake the feeling by thinking of something else, but her mind kept drifting back to that feeling that she was missing something. Perhaps it was simply knowing that she would be stuck sitting with a Niklas or two for who knows how long. She finished rinsing her body and worked the conditioner out of her hair and climbed out of the shower. She toweled off and slipped back into her robe. She laughed slightly, remembering how ridiculous it looked on Clay. She wrapped it a little more closely around her as she fixed her hair.

When she finished, she walked back into her bedroom and let the robe fall to the floor. As she stood there before her mirror she looked at her reflection. Her body was young and firm. Yet she could not help but feel a sense of sadness overcome her. A number of men had been pleased enough with her physical features to attempt to charm her by various methods. Several had been successful. But staring at herself that sense of loss revisited her again. She wondered whether her involvement in Leeds' group would ever offer the opportunity to settle down and stay in one place long enough to grow accustomed to the same body lying next to hers at night, the same arms gathered around her offering comfort and familiarity. She wrapped her own arms around her body and wondered if she would ever experience the feelings associated with a growing belly, the increase in weight, fatigue, perhaps some morning sickness, and later the engorgement of her breasts and the pains of labor, the creation of a new human life. She turned away and grabbed her phone. She dialed her father's number. There was no answer and she was transferred to his voice mail.

"Hi Dad. I just wanted to say that I'm leaving town for a little bit and won't be available. We've got some research we need to finish at the lab," she lied. She had no idea where she was headed. "I probably will be out of touch for a while, but I'll call when I can. Love you." She paused briefly. "Tell Mom that I love her, too." She hung up. She finished getting dressed and was just getting ready to clean up the mess from the previous evening when there was a knock on her door.

CHAPTER THREE

> *The President has announced today that the Bill to accept the admission of the Canadian Provinces and Territories into the Union will be signed later today in a formal ceremony in the Rose Garden. Citizens of the former county of Canada will begin receiving Genesis replicators within days. Inoculations of PreVentall shall be delayed several weeks until scientists complete steps necessary to distribute the Two-Oh upgrade. Government scientists have assured reporters that the inoculation of former Canadian citizens will not delay the release of Two-Oh to Americans already clamoring for access.*

Clay let out a heavy sigh as he rolled over in bed. Lillian was gone. She had left for the west coast to visit some friends. He did not know how long she would be gone. A month perhaps? Two? When she had announced the plans he had said that he would be unable to join her at first owing to the fact that he had made an appointment to get inoculated but they would not actually perform the procedure until Two-Oh was formally released. Since that date was uncertain, and since he did not want to miss the appointment, he would remain at home until he was inoculated. *I can always figure out an excuse later why I didn't,* he told himself at the time. Lillian was ecstatic when he surprised her with the news. Clay was equally happy. Although he had always enjoyed their trips out west, his growing disconnect with Lil always seemed worse when they were visiting friends. Being the 'freak' in the crowd was always a bit easier for Clay when he was home. *Comfort and familiarity*, he had convinced himself. But he knew better. He knew that it was easier on Lillian to enjoy herself when she wasn't with him. He stayed home for her, for what was left of them. When she had left, Clay had wished her a safe flight as she rose out of bed early that morning.

Clay listened to her as she showered and readied to go. As she returned to the bedroom, Clay, still laying in bed, had feigned sleep. She kissed him on the cheek and departed. They both played along with the charade that he had fallen back asleep, avoiding the awkward truth of their diverging lives. But then he really had fallen back to sleep and into that awful dream.

Clay sat up, not eager to face the day. Lilian had been gone for several weeks already, but in the past 12 hours, Clay had become convinced that today was the day he was going to be forced to face his deepest fears. He was convinced now, more than ever before, that inoculation was a grave mistake. *Yet what choice do I have?* he wondered. He stared at himself in the full length mirror standing in the corner of the room. His midsection protruded slightly from under sagging pectoral muscles. His arms seemed slighter than they had only a few years earlier, and any notions he held of simply ignoring the advice of the Niklases to keep his appointment were swept away by the reality of his position. *What can I really do?* Yet there had to be some method of avoiding inoculation. *Think, damn it!*

Thinking did not help. His mind wondered back to his nightmare of last night. He knew that missing his appointment was dangerous. The risk to those he loved was too great.

He dressed and as he reached for his iMeme he paused. It struck him that the simple oversight of wearing his iMeme could be an excuse for failing to arrive to his appointment in a timely fashion. He picked it up and tossed it into the basket in his room with his dirty clothes. *Wow, is it past two o'clock? Sorry, I guess the time got away from me today. I must have forgotten to put on my iMeme today. I'm not sure where it is.* It gave him little sense of freedom from the Niklases. The idea that he was, after all, an old man with a fading memory was a nice fantasy, but it would not be so easy. Perhaps he could manage to delay the appointment. Perhaps.

A hopeful thought crossed his mind. *Maybe Eva could help me devise a solution, a plan to delay the inoculation.* He wondered if she were awake yet. He considered calling her for a moment and then decided against it. *Don't call, it could be traceable.* It was all very much more complicated than he had hoped.

Back in the days before PreVentall, he started thinking, then stopped himself. *The days before . . . .* It sounded ominous to him. *The days before PreVentall. Time may no longer be referenced by the rise of Christianity – B.C. or A.D. – but the advent of the new world, the world of science over nature. B.P. and A.P, perhaps . . . .* He shook the idea from his head and his mind returned to thoughts of various television dramas in which the authorities were able to track a killer's movements by phone calls made on his cell phone, by determining which cell tower carried the particular calls. *No, it would be best if I can limit any way of tracing my actions back to Eva. At least, if I can.* Whatever he could do to avoid inoculation would certainly be risky, and he did not want to involve Eva anymore than she was already.

He wondered around the quiet house. It was not unusual for there to be no activity. He was not sure how Lillian generally spent her days, but it was not at home. Katie was off at school, although she was scheduled to be arriving home from college for the weekend. He could not recall any approaching holiday but, then again, he wasn't completely sure what day it was. Perhaps Katie was coming home simply to visit friends. *I'm not even sure I remember when she left,* he thought to himself. *And Matthew and Elizabeth? When was the last time I saw the two of them? When did we last eat dinner together?* His memory came up blank. A feeling of melancholy and loss came over him. He suddenly had the feeling that he was living outside the world. A shadow of a person moving among the living, present, yet not entirely there. *Is it possible that I can't remember when I last saw the kids?* He tried to shake the feeling. *Get a grip on yourself, Clay. You're just a little shaken up by the events over the past few days.* He saw a pile of Matthew's shoes near the back door, a dirty cereal bowl left in the sink. *How long had that been there?* Elizabeth somehow never managed to put her dishes back into the Genesis for disposal. He was not insane, simply preoccupied.

He slipped into his gym shoes and headed towards the front door. It looked to be another beautiful day out. He decided he would walk downtown. A little over nine miles. He could take his time getting downtown, slowly work his way towards Eva's apartment. It would give Eva a little more time to sleep in before he disturbed her, and give himself time to think whether it was even a good idea. He found himself instinctively looking toward the steps for the newspaper that had not been delivered in a long time. He was surprised to see an envelope lying off to the side on the lowest stair. *What the . . . ?* He bent down to pick it up. The mailman had, like everyone else, moved on to matters not involving his profession any longer. Of course, the mail industry had long since disappeared. One simply needed only to place whatever it was that they wanted to send into the Genesis, which would disassemble the item, record its molecular make-up and then recreate an exact duplicate in the Genesis of the recipient. A quick notification that one had a delivery would alert the recipient, who could then retrieve the sent item at his convenience.

He turned the envelope over in his hands. His name was scrawled, somewhat child-like, across the envelope on a peculiar slant. He felt a hard object slide towards the lower end of the envelope as he turned it. He ripped open one edge and the steel key dropped to the ground. He quickly retrieved it and pulled out a sheet of paper that had remained tucked inside the envelope.

> CLAY,
> HERE'S A KEY TO MY PLACE. I'M NOT SURE HOW LONG I'LL BE GONE. PLEASE BE CAREFUL AND TRY NOT TO LET ANYONE SEE YOU COMING OR GOING.
>
> EVA

Be careful? Clay thought about it for a moment. *Was she merely referring to stopping by her place, or did she mean something more? Was she suggesting I skip the appointment? That it could even be possible?* He looked toward the sun still low in the eastern sky. He slipped the key into his pocket and took a deep breath as he headed toward Jackson Boulevard. Jackson would take him to the Loop, and within a few blocks of Eva's place. It would also provide him the opportunity to walk through Columbus Park and Garfield Park, both large green spaces that even during the Great Depression II had retained their grandeur, though perhaps not their full majesty. Even in the early years of the economic collapse the Garfield Park

Conservatory had remained active, providing examples of exotic flora mixed with various other activities meant to attract persons not normally inclined to visit a large greenhouse: *Chocolate and Chihuly* for those with a sweet tooth or *Beer Under Glass*, for those with an affinity for micro-brews. The West Side was also filled with beautiful graystone houses that had, beginning in the days of White Flight, fallen into neglect and disrepair - and even complete devastation after the 1968 West Side riots following the assassination of Dr. Martin Luther King, Jr. – but had since regained their original glory with the use of InFIXstructure. Continuing east, he would pass what was once the Medical District, pass south of the United Center and into the Loop. He was looking forward to the hike. It would give him the opportunity to clear his head and think things through.

Clay crossed Austin Boulevard – the dividing line between affluent Oak Park and the once impoverished West Side neighborhood of Austin – and past the green lawns of Columbus Park. Thereafter, rows of greystone homes and large apartment buildings began to replace the greenery. The streets were relatively empty, but on the north side of the street Clay noticed several teenagers gathered around the front of one of the larger greystone homes. As he neared, one of the young men rose and shouted something while staring intently at Clay. For a brief moment Clay felt a surge of adrenalin pulse through his blood as a not-yet-forgotten instinct for fight or flight responded to the youth's seemingly threatening acts. A white man venturing alone through a predominately African-American and historically impoverished neighborhood. He paused for a moment and the young man shouted out again.

"What cha' doing here old man?" The teenager took a step towards Clay, but paused himself.

Clay recognized a certain hesitancy in the young man's actions and it struck him that the boy was not threatening Clay, but was, himself, threatened by the sight of an old man in a world of young people. Clay let out a short sigh and smiled back, waiving his arm as he spoke. "Good morning. Just taking a walk today. Enjoying the day." He continued his progress to the east.

The teen looked up and down the street and then crossed towards Clay. "Are you from around here? I mean . . . ." The boy stopped mid-sentence, unsure of what to say. His friends remained on the north side of the street.

"It's okay." Clay stopped walking for a moment. "Yes. I live over in Oak Park. I . . . I" it was Clay's turn to be at a loss for words. He paused, and then continued on. "I'm on my way to get inoculated and I thought I'd give this old body one last challenge before I change."

The youth gave him a strange look and it occurred to Clay that the notion of mortality, the leeriness of becoming inoculated, of displacing himself from humanity, would not have registered as very relevant to a teenager of pre-PreVentall days (*B.P* jumped into his thoughts) let alone one in a world of immortality.

"Would you come with me to my Granny?" the boy asked.

"What?" The question caught Clay completely by surprise.

The teen looked away briefly, then back to Clay. "My Granny says she don't need to be inoculated. Says she knows one who ain't got no shot and that she's waiting for him. You who she waitin' for?"

A puzzled look crossed Clay's face. He looked back towards the house the boy came from and saw a young boy and young girl leading an elderly woman out of the house, each holding firmly to one arm, more in an effort to pull her outside than to hold her up. The old woman, having cleared the entranceway, wiped the kids' grips from her arms and turned to look toward Clay. She was not a tall woman. Clay guessed she was not much taller than five feet, with brown skin the color of iced tea, and gray hair showing the traces of the black it most certainly had been when she was a younger woman. When she saw Clay a smile spread across her face, her white teeth flashing in the morning sun.

"Well, I'll be Chil'." She scurried across the street to meet Clay with a spring to her step Clay did not think possible at her age. "Why it's been so long." Her eyes flashed a look of recognition.

"Good morning," he started, extending out his hand towards the old woman. "I'm Clay—" but she interrupted him.

"Is that what they're calling you now?" she said, more to herself than to Clay. "I know who you are, Chil', I know who you are."

"And you are?" Clay ventured.

"I am. Come on now, you knows that. Sure as you be standing here, I am, Honey Chil'," she laughed.

"Granny Manayi," a young child on a bicycle called out, waiving his hand and continuing up the street, a smile stretched from ear to ear. Manayi took Clay's hand and clasped it tightly within hers, still smiling. "Do you have time for a glass of sweet tea?"

Clay was unsure of how to respond. *Who was this woman and how does she know me?* She tugged on his hands, leading him back across the street. Clay looked around again. The kids were all staring at the two of them, smiling. Clay had the distinct feeling that they were almost as uncertain of how to take all this as he.

"Sure," he responded at last. "I have time for some tea."

Clay entered the home and was surprised at what he saw. While the InFIXstructure had returned the exterior to its glorious past, the interior told of a less glorious period. While the home was neat and well-kept, it retained an element of poverty; cracked plaster and fading wallpaper. The furnishings were clean but simple and what it lacked in material comfort, Clay sensed, was made up for in love. Manayi led Clay into the kitchen, which was bathed in morning sunlight. A gentle breeze blew in from the open window. She motioned for Clay to sit at one end of a small table pushed up to the wall just below the window. At the other end stood an empty chair, one that Manayi would presumably occupy once she finished pouring the tea. Clay looked down upon the red and white plastic gingham tablecloth. In several places the vinyl had worn thin, exposing the flannel backing. The kitchen was small and clean. He glanced around but did not catch sight of a Genesis. *Could these folks have been forgotten?* he wondered.

"No Chil', we ain't been forgotten," Manayi said as she placed a tall glass of sweet tea on the table before Clay. "I don't want no Genesis in my kitchen. That ain't fit for eating nothing out of it." She sat down and smiled at Clay who was staring at her with his mouth slightly agape. "It's okay Chil'. I understand lots of things." She took a sip from her own glass.

Clay, not sure of what to say, took a sip from his glass. The tea was cold and crisp with a perfect balance between the bitterness of the tea and sweetness of the sugar. "Wow," he said. He took a long draught, nearly emptying the glass.

"I add a touch of orange peel when I'm steepin' the tea. It's an ole family secret."

Clay smiled at the woman. "It's delicious."

"'course it is," she said matter-of-factly. "I'll pour you another glass." She stood up and re-filled Clay's glass. "You gonna do it, huh?" she asked.

"What's that?" Clay questioned, looking at her more closely now.

"That inoculation thing. You gonna become one of them?"

Clay was silent. He hadn't really decided how the day would turn out. He took the glass she offered him from her hand and she sat back down. "What do you mean, 'them'?" he asked.

"Them folks who all thinks theys can live forever now," she said. "They can't. Ain't nothing can do that now. You know that, Chil'."

"I don't know what I know," he replied.

"Well let me tell you, Chil'. It don't make them like us. It don't leave you human no more either. It gets you stuck between.

"We was supposed to be one big family. Everything. Man, nature, all of us. But we done messed that up now, didn't we?" A look of conspiracy and loss clouded her face briefly. The two sat across each other in silence for some time, drinking their tea.

She reached across the table and patted Clay on his hand. "It's okay, Chil'. No one blaming you. But you think you learn something second time 'round." She stood up and removed the empty glasses from the table. "Too nice to be sittin' inside." She exited the kitchen.

Clay sat for several minutes before he realized she was not coming back. The house appeared to be empty of any human activity and Clay stepped out the front door. Manayi was standing at the foot of the porch stairs.

"I was wondering where you at, Clay." She started walking east down Jackson Street. Clay caught up to her.

"Where are you going?" he asked.

"It's not where we going. It's where we is. Where is you Chil'?" She continued gazing forward.

Clay was uncertain how to respond. "I'm right here, on Jackson."

"No, no Chil'!" There was a sternness in Manayi's voice. "Like the Good Lord done asked after eatin' of that apple, 'Where you at?'" She looked over at Clay. "Where you at, Clay?"

Clay became even more confused. "I'm sorry. I don't understand."

"Silly Chil'." Manayi chuckled and turned her head away. "Like He don't know where they be hiding?" she said softly, as if speaking to someone else.

"He done created the Heavens and the Earth and you think We don't know you people be hidin' your butts behind some trees?" Manayi let out another small chuckle, and Clay wondered if she were addressing him. She turned towards him. "He done know that Chil'. He be askin': 'Adam-chil', you know what you done did? You done know eatin' that apple gonna have consequences?'"

Clay stopped walking, his face crunched in non-understanding.

Manayi looked sternly at Clay again. Her voice turned dark with impatience. "He saying, 'You done eat the apple, now what you gonna do? You done made a choice - and a wrong choice at that Chil' - and now you gotta figure out how that choice gonna change your life.'"

"I . . . I don't know," Clay answered. He looked over at Manayi. She began walking, and he moved quickly to stay near. She was looking straight ahead again, paying little attention to Clay at the moment. "Do you know where you are?" he asked.

She stopped suddenly and turned to him. "I always know where I am Child." The hard "d" on the end of the word emphasizing her seriousness. "I's right here," and she chuckled and began walking once again.

"That's a little disingenuous now, isn't it?" Clay asked, jumping forward to catch up. "I mean, really? You think I should know where I am but slip out of answering where you are with an 'I'm right here'?" Clay looked around. "I guess I'm right here too."

Manayi stopped again and turned to Clay. The smile left her face and her eyes became hard. "You ain't be such a smart ass when the Lord be asking you," she said more to herself than to Clay. "I know where I am Chil'. I done always know where I am. When I say I right here, I mean just that. I be who I am and tomorrow I be who I was. I don't need no inoculation to know where I be. But you, Chil', I'm worried 'bout you. You done think you know where you is at but that can't be, given where you be going. Course, you never done have 'nuff sense to know where you is at."

"How do you know where I'm going?" Clay asked.

"I done see it in your eyes, Chil'. You whole body just about screaming it. You goin' down to become one of them." A dark look crossed her face.

"You don't understand. It's not my choice. What about my wife, my kids? Even Eva's in danger. If it was just me--"

"Just you? Where youse at Chil'? It ain't never just you." she interrupted. "Don't matters whats you do. If it was just you it wouldn't matter nothing nohow where you be going. Heck, Chil', the good Lord done ain't asked, 'Adam who you be?' Lord done know who we all are. But He keep asking where folks at. 'Adam, where you be at?' 'Cayin, where you brother Abel, be?' Don't you see Chil'? We all got to know if we doin' what be right or what be wrong. You be doin' what be right for youself? You allotted only so much time on this here world. You done know where you at? 'Right here,'" she snarled. "Question be whether you wanna be goin' where you be goin'."

Clay shook his head. He looked down slowly towards his feet. He was suddenly tired and felt the urge to sit down. He looked around and chose a soft patch of grass on the parkway. *What am I doing?* he thought, placing his head in this hands. He felt a pressure rising inside. He felt trapped, forced between two undesirable futures. *But what can I do? I can't run – there's nowhere to go. Even if there were, what would happen to Lill, to the kids?* He pondered a life on the run. He would be alone. No family, no friends, nothing. *Who could I talk to? Not even the Walking Man. He's certainly dead or imprisoned. Who were the others? Were they also captured? Does it matter? How could I find them anyway?* He let out a sigh.

Where am I? Clay pondered the question. *Maybe getting inoculated* is *the right thing to do. Maybe this belief that it is somehow unnatural is just my stupid paranoia. Why shouldn't I strive for eternal health? Spend an eternity with Lill and the kids? What kind of son-of-a-bitch walks away from his family? From his friends?* Clay felt the tightness in his chest, the pressure rising to avoid the wrong choice. *What is the right choice? What is the measure of my life?* He looked back towards where Manayi was standing, seeking her thoughts, but the sidewalk was empty. He turned to look all around, but the neighborhood was empty as far as he could see. He was further east than he had remembered being, nearly to the United Center. It was a relatively short walk to the Loop from here. *What time is it*? Silence. He remembered that he left his iMeme at home, leaving his inquiry unanswered.

Clay stood up slowly and began walking again, wondering if Manayi was real or a figment of his imagination. He had found that lately many things felt unreal. He had attributed it often to the inability to know the date, the inability to track the progress of his life. *An unnecessary task for the immortal, but,* it suddenly struck Clay, *integral to the very nature of Man. Didn't God, as His very first act of creation, separate the day and the night?* He looked up at the sun, trying to estimate the time of day. *But of course!* Clay thought. *How else would we know where we are?* A feeling of loss began to overtake Clay and he hurried his pace towards Eva's place.

* * *

The room was cool and quiet. Except for the mess from the previous night, he imagined that anyone entering the apartment could easily think it seldom inhabited. He picked up the empty vodka bottle lying sideways on the table near the couch. Two glasses sat amid rings of dried condensation on the table. Clay placed all these items into the sink and stepped over to the freezer. It was relatively empty, but did contain two unopened bottles of the homemade vodka he had tasted last night. *Was it last night?* He couldn't remember. He pulled one out. It chilled his fingers immediately as a thin layer of ice formed over those areas of the bottle not laying beneath his fingers. He unscrewed the top and drank straight from the bottle. It was cold and smooth. He set it down and walked around the apartment. The robe he wore the previous night sat on Eva's bedroom floor, the bed was unmade. By all appearances, Eva left in a hurry. Yet the notion of sudden departure could not explain how she managed to drop the key off at his house. A wave of exhaustion once again swept through his body. He was not prepared to get inoculated, but he was unable to formulate a plan to avoid it. *Think, damn it, think!* His mind drifted back to the bottle of vodka. He prepared to take another drink. *Christ! I've got to get out of here. What am I going to do? Just sit here and get drunk?* he thought. *I don't know what I'm going to do, but whatever happens, I can't just sit around here. Not here at Eva's place.* He put the vodka back in the freezer and pulled out the envelope with the key. He started to rummage through the kitchen drawers until he found the junk drawer. He took out a pen and tape. He wrote PRIVATE across the front of the envelope, trying to disguise his handwriting. He knew no one at home would recognize Eva's handwriting. He rummaged around and found note paper and carefully wrapped the key in the paper and then wrapped tape around it, winding it around until it formed a package unrecognizable by feel as a key. He placed the package in the envelope and sealed the torn edge of the envelope with tape. He took one last look around and left, the door locking behind him. *OK. No hiding at Eva's.*

Clay stepped out into the street and headed north towards the Loop. He passed a post box, or at least what passed as a post box these days. It was nothing more than a public Genesis. He pulled the envelope out of his pocket and paused. *Shit!* Without his iMeme, Clay could not program the post box to deliver the envelope to his house. He folded it back up and placed it in his pocket as he continued walking. He suddenly felt vulnerable. He was only a few blocks from Dr. Harmsfeld's office and he wondered what time it was. He looked around for any lighted display or other electronic clock, but saw none. As he looked around for someone to ask he thought he caught sight of a Niklas. He stopped suddenly. He instinctively turned his head away to avoid being seen. *Idiot! As if there are any other old men walking around the city.* But he felt he could not stop, so he continued walking, head down and unaware whether he was being followed or not. When he reached the river he walked west along the path to one of the many benches that lined the RiverWalk. He did not see sight of any Niklases. *God, am I tired,* he thought. He placed his hand on his thigh and felt the key in his pocket. *I can't get caught with this. If they found out it was Eva's . . . .* He was not interested in considering the ramifications.

Is there any way out of this? His thoughts passed briefly to his family and friends. *No, no,* he thought. *That just confuses everything. Let's simplify this by starting with just me.* Clay shook his head as if to clear his mind. *If I get inoculated then I become one of them.* He paused. *'Them'? Isn't that what Manayi called inoculated folks?* He wondered if he had ever verbalized, even to himself, the clear distinction he felt between himself and those who had been inoculated. *If I can avoid this, then what? A life on the run? What would happen to Lill and the kids? And how long do I really think I can go on?* The heaviness of exhaustion weighed on his shoulders. *But if I do it, they win. Humanity takes a final step towards extinction. Extinction,* he thought. *The end of an age.* He watched a couple running along the path in front of him. The question and Manayi's voice rang in his ears. *Where is you?*

Where am I? He contemplated his fate for a moment. *If I succumb, I kill myself by living forever. If I don't, I live by dying.* A cloud of confusion passed through his thoughts. He felt trapped. *A Catch 22,* he thought. The sun had moved further along in the sky. *I've got to figure this out.* He sank down onto the bench and closed his eyes. In a moment he was sound asleep.

CHAPTER FOUR

Attention travelers: Due to the severe outbreak of the Covid-19-B2 virus, all persons entering the country shall be subject to required temperature screening. All persons registering temperature higher than 37°C shall be subject to mandatory quarantine. Persons unable or unwilling to submit shall be refused entry into Israel.

It was Eva's first time at Ben Gurion Airport. She was surprised at the relatively small size but awed at the diversity and activity taking place, as well as the number of handsome dark-skinned Israeli soldiers among the crowd. She paused as she entered the terminal from a secure exit. The transport which carried her to Israel landed but rather than pulling into a gate, simply let the party off on the tarmac. A repeating announcement was being made over the airport public address system, which was first made in Hebrew and then English. *Covid-19-B2 outbreak? I hadn't heard anything about that - but then again, what do we hear about anything back in the States?* She looked to the left and right, seeking the location of a restroom. She spotted one and began heading for it when a Niklas grabbed her shoulder.

"This way," he said, sternly.

"Oh, I'm sorry. But if I don't get to the restroom, I'm going to have an issue."

Niklas scowled at her but released his grip on her shoulder. "Don't be too long," he warned.

Fuck! Couldn't I have been escorted in country by a handsome Israeli instead?

She took her time in the restroom, defying Niklas' admonition to be swift. She slowly washed her hands and studied her reflection in the mirror. She looked all the worse for the wear. It had been a long flight. It was then she noticed that another traveller was standing 6 feet behind her, awaiting her turn at the sink.

Covid, she thought. "I'm sorry. I . . . I . . ." Eva did not know what to say. "I'm sorry," she said again and exited the restroom.

Niklas was standing outside the bathroom entrance. "I wasn't sure you were coming back out."

"Oh, Niklas, what a joy to see you. I've missed you so dearly too," she said sarcastically.

Niklas leaned in a little toward Eva and whispered, "Professor, you are lucky that the Colonel likes you. Personally, I don't trust you. And if it were up to me," he said as he lightly punched his fist into his opposite palm.

"Lucky me then," she said, trying to sound confident. She shuttered at the thought of losing the good graces of Leeds. *Perhaps I really am lucky.*

Niklas stood a little taller and returned to the roll of dutiful soldier. "Please, this way Professor," he said as he began walking. Eva followed silently beside him. The line at Immigration and Passport Control was split between Israeli citizens and non-citizens. Eva slowed her pace, preparing to join the larger line of non-citizens through customs, but Niklas grabbed her elbow and nudged her toward what appeared to be an out-of-service counter, behind which sat a bored-looking Immigration Officer. She stood up as they approached. "Sergeant Mueller," he said as she nodded.

"This is Eva Diaz. She's with the American contingent."

"Passport, please," she asked flatly.

Eva began to reach into her bag, but before she had the chance, the Niklas presented both his and Eva's passports to the agent. Eva gave the Niklas a dirty look. *Ruffling through my bag on the flight over? I shouldn't have slept.* Eva had noticed that in the other lines, all people were stopped prior to approaching the Immigration Officer and scanned with contactless infrared thermometers. She noticed a couple two lines over that were tested and then stopped from advancing further. They were led from the line towards the back of the room, speaking loudly and clearly upset. Eva looked over her shoulder towards where the couple was being led and saw several signs in multiple languages bearing red lettering: QUARANTINE. *I guess they didn't pass,* she thought.

The immigration officer stamped Eva's passport and handed it back. Eva shot her arm toward the agent. "I'll take that, thanks," she said as she snatched her passport from the Niklas' waiting hand.

"You may proceed," the agent said as she gestured the two forward.

"But what about—" Eva started.

"We're Americans. We can't get Covid, or any other illness or disease. We're exempt. From everything," Niklas added. "Come along, Professor."

* * *

Bialik Street was bustling with activity. The presence of two Americans sitting at a small sidewalk table at *Cafe Bialik* would easily go unnoticed amongst the many European tourists and Israeli citizens whose ancestry passed through the West. Col. Leeds sat quietly, his cold blue eyes looking out at the passers-by, sipping a Nescafe. Eva had never considered how cosmopolitan Tel-Aviv was. Her cup of coffee sat beside her on the table growing cold.

"I'm sorry, my dear," Col. Leeds spoke, still staring at the crowds passing by. "I should have warned you that when you order a coffee here you are greeted with a cup of coffee served in the manner of the Turks, a thick black sludge hardly consumable by a human. The Turks have been making it that way for over 400 years. Savages." He leaned over but caught himself before he spit. As if the mere mention of anything he considered Arab brought a layer of filth into his mouth. A habit he developed from his days in Iraq.

Eva smiled at Leeds. It had been a long flight, strapped into a small folding seat in a Boeing C-17 transport, across from several Niklases that were also on board. She did not know her destination when she boarded, but given that the Niklases told her to bring her passport when they picked her up, she knew that it must be outside the United States. In a normal flight, Eva generally liked to sit in the aisle row. Any interest in looking out the window of an airplane dissolved many years earlier into the reality of flat views of an expanse of ocean or the darkness of night. The aisle row also provided the opportunity of ease in getting up to stretch or use the restroom. Whether or not the C-17 had a lavatory was something Eva never discovered. She had feigned sleep to avoid the monotony of staring into the Niklas' expressionless faces. A tense and bumpy flight that seemed much longer than the twelve plus hours it actually took.

But now, sitting in the sunshine on the streets of Tel Aviv, she felt relaxed. The visual pleasure of being surrounded by dozens of people of various ages brought her a sense of normalcy. The hustle and bustle of activity: soldiers traveling here and there, business persons heading to somewhere, young people in groups laughing and talking. She noticed the streets were generally clean, only marred by large posters bearing the image of what appeared to be large splash of red paint with a black symbol in the middle, a sort of half square, the lower edge smooth and straight and the right side curving into the top line, that ended with a swoop. A single dot occupied the middle. It reminded Eva of a face of sorts, a collared man staring blankly with his single eye, the left half of his features missing. The posters also displayed Hebrew lettering above and below, scattered randomly on lamp posts and occasionally the side of a building. She stared intently at one. *Perhaps an advertisement for a concert or festival,* she thought, but returned her attention to Colonel Leeds.

"No apologies needed, Colonel," Eva said. "I don't think I would drink it if it were the finest cup of java in the world. I'm just content to sit here and take in the view." She turned to look at Leeds, who continued to stare out to the street. She wondered whether he even saw the activity on the street, or whether he was simply directing his gaze outward, contemplating other matters. He seemed to take little notice of the many people surrounding them and barely acknowledged a Niklas as he stood up from the table behind them and whispered into Leeds' ear, other than to raise his hand briefly, his index finger pointing straight up. The Niklas turned briefly toward Eva and gave her a stern look before returning to his seat at the table behind the two.

Leeds turned toward Eva, his eyes focusing on hers. "It seems we have a small situation at hand. I'd hate to break up your pleasantries here, but this must be attended to immediately." Leeds stood up and held out his hand. "Please, my dear Professor. We shall go now."

Eva instinctively reached out to Leeds and took his hand as he helped her rise from her chair. These little acts of humanity took Eva by surprise almost as much as the cold, clammy skin that always greeted her fingers when her hand met his. Leeds puzzled Eva. He seemed so confident and sure, yet she sensed a hint of nervousness on the several occasions when he offered her his hand to assist her or as he reached out to hold hers prior to kissing it. These formalities seemed befitting of an officer, yet it seemed more than that. It was during these brief moments that Eva sensed something beyond the cold exterior that normally was on display, as if Leeds were actually capable of deeper human emotion. She struggled to rectify these moments with his general demeanor and sometimes had the uneasy feeling that Leeds was attracted to her; that he wanted to be more than simply her boss. Perhaps a friend. Perhaps more than a friend.

She smiled at Leeds as she pulled her hand away. "Thank you." She brushed her hands down her hips and thighs, straightening her skirt, wiping away the trace of clammy moisture that lingered on her fingers from having held Leeds' hand, as well as the idea of an intimate friendship between the two. He was a powerful man and she wondered whether her niceties towards him were as much out of fear and caution as friendship. She did not hate the man, like everyone else she knew who worked with him. Everyone except the Niklases, who seemed to be almost infatuated with him. *A man-crush,* she thought. Leeds was a cruel and seemingly heartless individual, traits she was certain the Niklases shared and admired, yet he always treated Eva with kindness and respect. *Why is it?* she wondered. She struggled with her own feelings towards the man. She recalled the many occasions the two shared a laugh and enjoyed each other's company, yet her gut told her that she should not trust him, or even call him a friend. She knew, however, she could not break away from his power over her. He was too dangerous to turn her back on, and, for all she knew, was a pawn to a bigger and more powerful player. She had long ago decided she may need him as an ally. She was certain she did not need him as an enemy.

Eva and Leeds sat in the back seat of the limousine alone. Two Niklases sat up front, hidden behind smoked glass. Eva was certain that their conversation could be private, if Leeds had wanted it to be. But she was unsure what he wanted the Niklases to know or not and made a mental note to herself to watch what she said.

"Would you care for anything, my dear Professor?" Leeds asked as he opened the chest cooler, revealing bottled water, soda and various fruit drinks. He paused a moment. "Or perhaps something a little more interesting." He slid a panel open to reveal a full wet bar as he looked at his watch. "It helps us that we both wear watches," he said looking at Eva's wrist and smiling. "An American may not even remember what a watch is."

She smiled back. The cold air blowing out of the vent reminded her of America and she longed to open the window. To feel the heat of the desert now rising around them to fill the car with discomfort and sand; to remind her what it is to live among the imperfections of a world that no longer existed back home. She depressed the button to roll her window down but the glass remained steadfastly in place.

"I'm sorry my dear. I have no tolerance for the desert. If it were not for Israel, I would never set foot in a desert again. Not even the American desert. But our task brings us here, and so I ask that you suffer me my peculiarities."

"What are we doing here, Colonel?" Eva ventured, hoping to learn something of her presence.

"I am here for many reasons. Most of which you need not worry about. But you are here because I need you."

"You always say that," Eva smiled, trying to assess the situation.

"But it's true. It was through your discoveries that we were able to replicate sentient beings. Your success literally surrounds us," Leeds smiled and glanced toward the dark glass separating them from the two Niklases sitting in the front of the limousine. "But I have been thinking that we can do better."

"Better?" Eva asked. "In what sense?"

Leeds smiled and turned to look out the window. "Always about business. Did you notice the cactus along the side of the road? The ones with the red growths? You may know them as prickly pears, but they're called Sabras here. They are not native to this land, but have acclimated quite well. They are fierce plants, resilient to the forces of nature in this desert clime, yet contain a delicate sweet inside. Did you know that is what they call the native Israelis?"

"What's that?" Eva asked confused.

"Sabras. They call people who are born in this country Sabras. They, too, are a fierce and resilient bunch. But said to be sweet as well, once you get to know them."

Leeds continued to look out the window and the two rode in silence for quite some time. Eva had watched the landscape change from bustling city to arid desert, the brownish land sweeping past the car interrupted in frequent intervals by the sight of lush green farms and pastures rising up out of the desert like some fabled oasis in a tale from the *1001 Nights*. Eva was amazed at the ingenuity of the Israelis in turning the barren wasteland of the desert into a habitable and productive landscape.

"It is mostly drip-irrigation," Leeds said, breaking the silence. "Ingenious really. They can control the rate of drip and the hoses are all buried to avoid evaporation. Every precious drop of water used precisely for its intended purpose."

"Necessity really is the mother of invention then," Eva said, not sure what else to say.

"But of course," Leeds responded. "It is by necessity that I've come to bring you here," Leeds continued as the car turned a corner. Eva noticed a peeling poster, bearing the same spot and face design that she had seen in Tel Aviv attached to the driveway gate as they pulled past and up the drive. *Hmmm,* she wondered.

"Ahhh. We are here." Leeds opened the door and stepped outside, while one of the Niklases opened Eva's door. "This way, my dear," Leeds continued.

A petite dark haired woman wearing a tight pink dress came bounding out of the door of the home and immediately grabbed Leeds' hands. "Oh, Colonel. You are back!" she squealed as she bounced on her toes. As Eva walked around the car towards the front entrance she noticed that the woman appeared older than Eva first believed. From a distance, given the clothing and mannerism, Eva would have guessed the woman was in her late twenties or early thirties, but upon closer examination Eva could see that she was much closer to sixty than to twenty. "I have been longing for your return, Colonel," the woman continued, tears beginning to run down her cheeks.

"We have company, Nawra. Do not forget your manners," Leeds said abruptly, pulling his hands from her. "Please see that she is made comfortable. Afterwards we can see to your needs." Leeds turned and entered the home without any word to either woman.

"I am sorry, Miss," Nawra said. "I will help you. Please. Where are your baggage? I will show you your room."

Eva stood awkwardly for a moment. She was not told where she was headed and had nothing more than the small bag over her shoulder. "This is it," she said, raising the shoulder with the bag on it. "I guess I'll have to do some shopping while I'm here," she said, wondering momentarily how long that may be.

"Okay," Nawra said. "You will follow me, please."

As Eva followed Nawra into the home and up a flight of stairs, she suddenly felt tired. The artificially cool air had robbed her of the energy and excitement of being on the streets of Tel Aviv. Nawra opened the door to a large bedroom.

"You will stay here. There is restroom there," Nawra said pointing towards two open doors revealing a large bathroom area with a glass shower, soaker tub, and a large marble his-and-hers sink with an expansive mirror overhead.

Eva tossed her bag onto the bed, surprised at how much effort it took her to do so. She felt the need to lie down. "Thank you, Nawra. I'm just going to, to-" she sat down on the bed and quickly lay back, enjoying the soft comfort and let out a soft moan.

"I will leave you, Miss," Nawra said and exited the room.

* * *

Eva awoke with a start. She looked around for a moment at the unfamiliar surroundings and sat up in a confused panic. She let out a sigh of relief and relaxed as she remembered where she was.

Sunlight no longer poured through the windows and she looked down at her wrist. Her watch read 5:17. She had not bothered to set her watch earlier and she could not remember how many hours ahead of Chicago Tel Aviv was time-wise. She guessed seven or eight hours, making it sometime past midnight. She raised herself from the bed and walked into the bathroom, thankful she had a change of clothes and a toothbrush in her bag. As the warm water of the shower massaged her skin she was reminded of the warmth of the air earlier in the day, and the unexplained joy she felt at being outside among hundreds of non-inoculated people. As she dressed and finished getting ready she decided that she would go out and explore the area. Tel Aviv was certainly cosmopolitan enough and she hoped that wherever she was now would not be too different. Perhaps she would find something entertaining no matter the hour. She stepped out of the room into the lighted hallway. She was touched by the thought that someone left the light on. She looked down the hall. The doors were shut on all the other rooms. *Probably all bedrooms*, she thought. She made her way down the stairs which ended across from the door through which she had entered the home. To her left was a darkened room. She could see book-lined shelves from the light spilling in from the hallway. The room to her right was dimly lit from the glow of a large television, the murmur of man and a woman on-screen having a serious moment barely audible. As the camera cut back and forth between close ups of each actor, Eva was reminded of the many American soap operas that used to dominate day-time network television. She did not need to speak Hebrew or have the volume raised to know that one of these two on-screen lovers were likely revealing the identity of the third person involved in a love-triangle.

Another hallway continued down to the right and Eva saw lights coming from the two doorways at the end. As she neared she could hear the quiet laughter of a woman.

"Professor. I trust you had nice nap?" Leeds asked as Nawra slid off his lap and eyed Eva, her lips still curled in a smile.

"Thank you, Colonel. I did," she responded, not taking her eyes off of Nawra who now looked positively radiant, her earlier vestiges of middle age no longer present.

Leeds raised his hand and flicked it, shooing Nawra away. "Enough for now, Nawra. Professor Diaz and I have business to discuss." He turned to Eva. "I will have Nawra bring you something to eat. I am certain you must be famished."

Nawra's smile turned into a sneer, but it quickly snapped back as she turned her gaze back to Leeds. "This time. But I will not be her servant. I save that for you." She took a cigarette from the ashtray and brushed past Eva.

Eva took a look around. The dining room was dominated by a large table capable of seating a dozen people. Artwork hung on the walls on one side. The other had four large windows, the curtains drawn on all, preventing Eva from seeing what lay beyond. There was a doorway to the right of Eva, which led past a butler's bar and into the kitchen. Another doorway stood across, the door closed.

"Never mind her, Professor. She is a wicked woman, but oh so loyal. Not to mention she makes the most glorious lamb Tajine with mango. The first time she made it, she was still living in Morocco. She used that rancid native beast the Moroccan's call lamb. It's generally hardly even palatable. But Nawra is magic, I tell you. Truly divine. I will be certain she treats you to some. And with proper lamb . . . well, I can't even begin to describe it."

"Magic?" Eva asked.

"Oh, Professor. Even you should not be surprised by her change in appearance." Leeds said, expressing some disappointment at the questioning expression crossing Eva's face.

"But-" Eva started.

"Don't be unreasonable, Professor. I have not violated Federal restrictions by procuring PreVentall and inoculating Nawra since your arrival." A wry smile crossed Leeds' lips. "While the promise of doing so can work to one's advantage for the short-term, I rather need Nawra for some, shall we say, long-term plans." Leeds paused and his smile widened. "I find that in such instances, the use of ReGenbots provides *continued motivation* for cooperation."

ReGenbots had been developed as part of the early PreVentall research for military applications. The Department of Defense issued a report which found that many soldiers were dying from loss of blood from wounds suffered on the battlefield. DoD further found that most soldiers could have survived their trauma if medical treatment in the field could limit blood loss. ReGenbots were simpler than the bots used in PreVentall, having no ability to self-replicate and no ability to migrate much beyond the epithelial layer of one's skin in the area to which they were applied, thus making only surface defects susceptible to correction. But when applied to traumatic injuries in the battlefield, they repaired open wounds and stopped the bleeding, as well as prevented infection. Eva was aware that ReGenbot production had mostly ceased with the introduction of PreVentall, but limited production continued, mostly for supplying to American allies around the world, though other uses continued to exist as well. *There's a use that never crossed my mind,* Eva thought, still staring at Leeds.

Nawra returned with a plate and set it before Eva. "It is late. You should not eat heavy."

Eva looked at the plate. It was arranged with olives, grapes, pomegranate and what appeared to be a soft cheese. Nawra set down a basket of bread as well. "Thank you," she said, looking up at Nawra.

"I do it for him," she said, smiling at Leeds. She gave Eva a small sneer before turning away and leaving once again.

Leeds leaned back in his chair. "You can ignore what Nawra has to say. She does this for neither I nor you, but only for herself." He reached over and took several olives from the plate. "I have made it abundantly clear to her that if she hopes to continue to receive the age-defying benefits of the ReGenbots I allow, she must treat you with the same deference as she treats me. She is a vain woman. You will find her quite willing to cater to any request you may have."

Eva said nothing.

"You, however, are a different story." Leeds' face lightened and the hint of a smile crossed his lips. "I can only hope that you will allow me the privilege of your intelligence and that you, dear Professor, will cater to my humble request."

"And what would that be, Colonel?" Eva ventured. "You know I want nothing more than to see you happy," she said, hoping she had hit the right tone between the sincerity of truly wanting to help the Colonel and a facetious, but good-natured, bite of letting him know that her feelings towards him were less dear than his toward she.

A full grin swept across Leeds' face. "Now, now, Eva. You know I would not have pulled you out of bed and dragged you half-way across the world unless I really did need your help." His tone had softened. "And I really do. It's these damn Niklases."

"Damn Niklases?" Eva's eyes widened in mock surprise. "Now you've got me intrigued," she said, placing some cheese on a slice of bread.

"Oh, don't get excited, Professor. My admiration for the Niklases has not waned. But I do find myself continually having to repeat myself." Leeds stood and walked over to butler bar, took down one of the large wine glasses and poured from a bottle that had been sitting on the counter. "Chateau Golan Gesham, vintage 2019. Another reason the Israelis annexed the Golan Heights. *Wine Spectator* rated it a 96. Quite exquisite for Israel, really." Leeds handed the glass to Eva.

She swirled the glass and took three quick sniffs. "Pomegranate , or maybe cherry. Pepper," she said. She raised the glass to her lips. She tasted sweetness on the front followed by acidity and a dry finish. "Oh, that is nice, Colonel," she said sincerely.

Leeds smiled briefly as he sat down, acknowledging her compliment. "Yes. But back to my dilemma." He reached over and pressed the button on a small electronic device sitting on the table. Eva had not noticed it before. Almost immediately a Niklas entered the room and stood at attention before Leeds. "Sgt. Mueller, I want you to head out to site Tango Charlie to pick up a package from Rav Seren Shalid. It is imparative that you return before Oh-Three-Thirty hours."

"Yes, Sir," the Niklas said, turning to go.

"Oh, and Sergeant?" Leeds said.

"Yes, Colonel?"

"I've just made a hefty wager with the Professor here that I could refrain from smoking in her presence for the next twenty-four hours. Please dispose of these for me," Leeds said, offering his pack of Marlboros to Niklas. "I'd be quite upset to lose this particular bet and I don't need the temptation."

"Yes, Sir. Understood, Sir," Niklas said as he departed the room.

Leeds smiled and Eva but said nothing. She opened her mouth to speak, but the Colonel held up his finger and motioned for her to remain silent. Eva stared blankly at the curtain drawn-windows, wondering what may lay beyond in the darkness. Her thoughts swiftly returned to the hustle and bustle of Tel Aviv, the throngs of people, both old and young, living their lives in a world so different than what America had come to know. After a few moments Eva could hear the sound of a vehicle moving across the driveway and disappearing into the night. Leeds once again depressed the button on the small electronic device on the table. Like the previous time, a Niklas soon appeared in the doorway.

"Sergeant, I seem to have misplaced my cigarettes. Would you please be so kind as to lend me a smoke? The good Professor here feels quite strongly against the habit, but I simply must insist," Leeds said looking at the Niklas.

"Why certainly, Sir." The Niklas reached into his pocket and retrieved a pack of cigarettes. "Please, Sir, have the whole pack," the Niklas offered.

"Oh no. I couldn't," Leeds said, belying his words and placing the package of cigarettes in his breast pocket..

"No, Sir. I insist," the Niklas said.

The room returned to silence for several moments before the Niklas spoke. "If it please you, Sir," he said placing his hands behind his back, "if you have nothing further, I have some duties to attend to."

"No, no, of course not, Sergeant. Please, do carry on," Leeds offered.

The Niklas turned and left Leeds and Eva to themselves. Leeds pulled a cigarette from the pack. "This, my dear Professor, is a bet I can afford to lose. There are others I cannot." He gave Eva a hard look. "Tell me, Professor, what good is it to me to have a conversation with one Niklas and then have to re-explain myself to another Niklas later?"

"I don't know. But I don't see how it is any different than having to re-communicate anything you may share with one person to another."

"Ahh, but there is where you are failing in your thought process." Leeds reached for a lighter, but then thought twice and placed it back down. "Perhaps it is because you have no affinity for my dear Niklases and you tend to keep your distance. You should know, dear Doctor, that they are fiercely loyal and I would never forgive them if they harmed you in any way. But all that aside, your lack of interaction has, perhaps, caused you to overlook the obvious."

Eva thought for a moment, but her mind kept wandering back to the outdoors, to throngs of foreigners enjoying the simplicity of living. "I give up," she signed, hoping that Leeds would not be angry for her inability to solve his riddle.

"There are too many Niklases. I need only one."

A puzzled expression crossed Eva's face. "Excuse me?"

"Don't get me wrong. I need each and every Niklas that I have. But, if you dealt with them more often, you'd realize that you'd need to be able to tell them apart, or better yet, not to." Eva's countenance continued to convey puzzlement. Leeds went on. "To whom have I spoken? To which Niklas did I provide certain information? Since they cannot be told apart, they need to be identical. I need to be able to speak to one Niklas and know that each and every Niklas knows what was said and what was expected, as if each were the only Niklas with whom I communicated, or more accurately, as if there were only one Niklas in the world."

"So, you need some type of universal communication system?" Eva ventured.

Leeds let out a laugh. "No, my dear. That is much too simple for the likes of you. Any number of scientists could solve such a simple riddle as that. What I need is a shared consciousness. I need each thought, to be shared by all, each synapses occurring in any one of the Niklases to be developed in each. To truly create duplicates of Niklas. Not only from the moment of replication, but moving forward in real time."

Eva stared silently at Leeds, forgetting for the moment about the world outside.

"And we need to complete our work quickly. We have very little time."

At that moment, the sound of glasses lightly tapping one another came from the butler's closet. Leeds turned to see Nawra standing in the shadows. "Nawra, you wicked wench!" he called to her as he stood. She turned to retreat to the kitchen but he grabbed her arm and pulled her into the dining room. "My precious little Jezabel," he went on. "Yes, you have finally learned of my secrets," he smiled. "Now, I no longer can control you. I concede that my powers over you have been diminished. You shall get your desire, if only to protect my secrets."

Nawra smiled, her eyes filled with the sparkle of victory. "Colonel, I know you cannot hurt me. I do not wish to harm you. But I must think of myself. Your secret is safe as long as I am young. But if I am old, I cannot perhaps help myself." She, too, smiled.

"Then it is so," Leeds said, reaching into his pocket and pulling out a small file. He poured the contents into his glass of wine and handed it to Nawra. "PreVentall, you wicked woman." He paused for a moment. "Your beauty shall reign until the Jordan River runs dry."

Nawra grabbed the glass and quickly swallowed its contents. "You're secret is safe with me Colonel. Forever."

"Of that I am certain," Leeds said as Nawra slumped to the floor. Eva jumped to her feet but Leeds put his hand up. He stared at her lifeless body but made no move to assist her. "Do not fear, Eva. It is but one of several incidents she has been made to suffer. She is fine, I assure you." He pulled another small vial out of his pocket and handed it to Eva. "I have found that it is easier to protect myself when others are made to forget."

"I don't understand," Eva said returning to her seat and examining the small vial.

"Forget-Me-Bots." Leeds curled his lips in feigned disgust. "A horrible name, but the lab was simply not up to its usual creativity when these were developed. You can give these to anyone, inoculated or not. These bots will interfere with the brain's ability to create the proper synapsis, thus effectively erasing an individual's memory of all events occurring over the previous 6-8 hours." He looked down at Nawra's still motionless body. "She will come to in about an hour, having forgotten not only what she may have overheard of our conversation, but of her indiscretions as well. This will save me much time. And her much pain."

Eva sat speechless.

"You are no longer in the States, my dear. You may find that these will come in handy to you someday." He saw that Eva was still unconvinced. "You can thank me later." Leeds beckoned another Niklas.

"Sir," the Niklas said before looking down upon Nawra's body lying on the floor. Without another word the Niklas lifted Nawra off the floor, slung her over his shoulder and retreated towards the kitchen.

"We can begin to resolve my problem in the morning. It is late and you may desire a little more sleep before the work begins. Shall I have Niklas escort you back to your room?"

"Thank you Colonel. I believe I can find my own way."

As Eva stood to leave Leeds also rose from his chair. He bowed his head slightly. "Then I bid you goodnight, Professor."

"Good night Colonel," Eva said as she turned and walked back towards her bedroom, still holding the small vial in her hand.

CHAPTER FIVE

Where am I?

Clay sat upright quickly. He had thought only moments ago that he was lying in his own bed, but the setting was all wrong. Where his dresser would normally sit was a long, low table, lined with a row of canisters of some sort. And from where the window across from his side of the bed would normally allow a dull glow of the street light outside to seep in from around the blinds was, instead, a solid form, broken by a poster, only partially visible in the dim light, of what Clay could only guess to be an anatomical representation of a human body. He turned his head to the left and nearly let out a scream, but the calming voice with a vague familiarity stopped the sound in his throat.

"Have we had a good nap, Chil'?" Manayi was sitting in a chair in the far corner of the room. She was wearing a white dress, with a zipper down the front, white stockings and shoes.

"What? Where am I?" Clay said, scanning the room once again, seeking some source of recollection, some glimmer of memory on how he came to be there.

"You see, now you be coming 'round," Manayi said, then softly to herself, smiling, "Where am I? Now ain't that funny? Clay be asking *me* where he be at." Manayi lifted her head up and stared into Clay's eyes. It was a hard stare, all traces of her brief pleasure gone. "Ain't that what I asking you all along? Where is you Chil'?"

Clay opened his mouth to respond but he was interrupted by the sound of a doorknob turning. The bright light glaring in from the hall preventing Clay from discerning anything other than the outline of a tall figure entering the room. The individual bent over towards Manayi, and Clay heard the murmur of a deep voice whispering, but was unable to make out anything of what was being said. Clay's eyes had still not readjusted to the dark and as quickly as the figure entered the room, it left in another blinding flash of light from the hallway.

"He say this time you get to choose," she said.
Clay sat quietly, waiting for Manayi to continue. But she said nothing further.

"I'm sorry," Clay said, unsure of anything. "I don't understand."

"Oh, Honey-Chil'. You asking a woman to decide whether you be doin' right or doin' wrong? Who get the blame you choose wrong?" Manayi shook her head. "Always be the woman's fault, don't it?

Manayi's pause was brief. "We ain't got much time. You need be deciding whether you gonna choose to be one of 'them' or not. Them Repeats know you here. They jus' too damn dumb to know he be here, too. He kin get you out, Child, iffin you want out. You wanna be one of *them,* then he and me, we's just leave an' that be that."

Clay sat for a moment, trying to absorb what she just said.

"Ain't got much time, Chil'. Where you at?"

Clay moved to stand, but then stopped himself. *This is it, then,* he thought. The enormity of his decision weighed on his mind. Every muscle in his body urged him to stand and shout, *Let's go!* and he stood ready to heed the desire of his human soul, but he paused. "What about Lill? The kids? Will I— " he paused for a moment, thinking of his family. "Will I ever see them again?"

Manayi shook her head slowly. "Don' know, Chil', don' know. We all sees each other 'gain, I suppose, but 'them' ain't be one of us no more. Don' know what happens to 'em." She sat quietly and Clay understood her silence to be deep contemplation.

She turned quickly back to Clay and broke the silence again. "Guessin' ain't no way for 'them' to know where they's at. How'n we supposed to see 'em again if'n they don't even know where they's at? But that ain't our problem, Chil'. That be somethin' between 'them' and Him. Everyone's gotta know where they's at. So where you at, Child?"

But at that moment, Clay was unsure. The desire to avoid PreVentall, the urge to be true to himself, to live and die as a man was overwhelming. *But how can I simply leave? Can I give up seeing Lill again, or Katie or Matthew or Lizzie?* The absurdity of the situation struck Clay suddenly. "I choose death. Death as a man, a real man," he said aloud.

Manayi smiled at Clay. She looked into his eyes, catching the flash of humanity that shone from them.

To refuse PreVentall is to choose death, but merely of my mortal self. If man is truly born to die, then refusing PreVentall is the last great act of humanity, Clay thought.

'Today we declare you Man, Mr. Furstman.' Clay said softly, more to himself than Manayi. The thoughts of never seeing his family again were tempered by the fact that his pain would be limited by his mortality. He thought of Lill, her beauty now eternal. He thought of his children. Lizzie was still a child. He would miss her transformation to full womanhood. Matthew, whose wit and humor would always keep him surrounded by friends. And Katie.

Manayi's smile quickly disappeared as Clay's body stiffened and the light faded from his eyes. Clay's thoughts were consumed by images of Katie transformed into the humanoid monster of his nightmares.

"Chil', we best be going from here." Manayi's voice interrupted Clay's thoughts. "Ain't got much time." She could sense the change in Clay. "You sure's took a long time to figure out where you's at."

Clay stood up, ready for whatever may come, but he found his strength leaving him. He felt nauseous at the thought of being inoculated; of becoming one of them. But his thoughts returned to his nightmare. *This is all wrong,* he thought. *But the alternative is worse. I can choose death for myself. But not for them.* He shook his head, trying to clear his thoughts.

Manayi had stepped back, shaking her head, sobbing softly to herself. "Oh Chil'. You knows where's you at. 'Them' folks is not right. You gots to hold on to where you's at."

Clay shook his head and then turned away, tears coming to his eyes. He could not bear to see her leave but he could not help but hear the click of the door as it was quietly pulled shut.

The nausea returned and Clay looked around. He did not feel he had the strength to lift himself to the sink so he spat on the floor. His mouth had begun to fill with the excessive amounts of saliva that always preceded the emptying of his stomach's contents.

Clay raised his head to the sound of the door being abruptly opened. Two men, shorter than the last man who entered, but stockier, more threatening, entered the room.

"Mr. Furstman?" the first Niklas asked, frowning at the pool of vomit beside the table.

Clay raised his head slightly and managed a weak response. "That'd be me," he said, before turning toward the floor again.

"It's your lucky day, BAMF. By tomorrow morning you won't be an old fuck any longer," the first Niklas said.

"I thought for a moment you weren't going to show today," said the second, his face morphing into a sneer. "I don't trust your type; avoiding everything we've created to make your life better. What's wrong with you?" The second Niklas paused. The first Niklas, did not wait for Clay to answer before responding. "I think you're a fucking insurgent." He turned towards his fellow self. "What do you say we kill him instead? We can say he tried to escape and we had to kill him." A smile grew on both the faces of the Niklases, as if Clay were watching a single individual smile at himself in a mirror. The Niklas on the left pulled out a small black device from his belt and moved toward Clay, nodding at the other Niklas. Clay stood up clumsily and slipped on his own vomit. He bumped noisily into a table and stood a little more upright. One of the Niklases grabbed Clay, squeezing his arm tightly when a voice came from behind.

"Ahem." The Nicklases turned around at the same time, the first releasing his grip on Clay. Manayi stood there, her white dress now accompanied by a cap. *A nurse's uniform. How odd,* Clay thought, unsure of what would happen next.

The left Niklas leaned in close to Clay. "Did you know that the doctors like to insure the shot is effective," he whispered. The right Niklas continued, "Usually they simply cut your finger to confirm it heals, but Doctor said we can run our own test. Let's say that our test is a little more prejudicial. Of course, there won't be any scars or lasting disfigurement, and any pain," he paused. The left Niklas finished the right Niklas' sentence, "should be only instantaneous, unless, of course, the nanobots take longer to activate then expected." Both Niklases laughed.

"Doctor says she ready for the patient," Manayi said looking straight at Clay. "You fellows gots to leave," she said, turning to the Niklases and effectively pushing them from the room. Clay was amazed at her apparent strength. One of the Niklases opened his mouth in protest, but Manayi stopped him.

"Don't make me get the Lieutenant in here. Now you boys git out. Shoo!" she said as she pushed the first past the threshhold and impressing the other Niklas enough with her tenacity that he shrugged to the other as he left the room, dragging his feet slightly as if he were a small boy whose mother had just kicked him out of the kitchen for trying to steal cookies.

She turned her attention back to Clay. "Now you jus' sit down, Chil'. You done made up your mind, I su'pose. I hope you know what youse doing." Manayi opened a drawer from the table Clay had just bumped into and pulled a small silver tray with a syringe lying upon it out and set it on the desk. She looked briefly over her shoulder and then removed the syringe and replaced it with one she had in the front pocket of her uniform. Then she carefully placed the tray back into the drawer and slid it shut. Clay watched her every move intently.

"I want you to understand, Manayi," Clay said apologetically. "I can't allow harm to come to my family. You understand that, don't you?" Clay asked, seeking her approval.

"Ain't nothing we can do 'bout it now. Doctor'll be in in a moment. You done made your decision, Honey-Chil'." She pressed against Clay's shoulders and he sat on the exam table. Manayi gave him a tender look, staring deep into his eyes. "But it okay. He give me this here. He say it version one, or somethin', and it be better 'n you havin' version two. He hopin' he can fix you up later. Talkin' bout some crazy somethin' or 'nother. I don't know 'bout no science stuff or nothin'. I just know you take care yourself and you keep rememberin' where you's at. Long as you do that, they can't take nothin' from you that's worth nothin'." Manayi gave each of Clay's hand, in turn, a gentle squeeze while continuing to stare directly into his eyes. Clay had returned the gaze and noticed that hers were a light brown, deep and seemingly full of their own source of illumination. He was transfixed at their beauty and was overcome by a sense of familiarity, of recognition, that went beyond their meeting earlier in the day. He closed his eyes, trying to find the connection as Manayi walked towards the door. As she approached the threshhold she turned back to Clay.

"Next time He asks where you's at Child, you go on and answer Him." Clay opened his eyes, her voice distracting his thoughts. "Don' go makin' no 'scuses for yourself. You just stand tall and lets Him know where you's at." With that she walked through the threshhold and was gone.

Clay closed his eyes once again, trying to rekindle his memory, but was once again interrupted.

"Mr. Furstman?" Dr. Harmsfeld asked as she entered the room. Two Niklases followed closely behind. She looked Clay up and down. "A holdout, huh?" she said wearily. "Maybe a good move. You'll be one of the first civilians to get version Two point Oh. Congrats," she said as she removed the syringe from the drawer and lifted it.

No!, Clay thought. He shifted his weight to stand up to leave, but he was restrained by the Niklases holding him him place. "No!" Clay shouted out as Dr. Harmsfeld plunged the syringe deep into his vein.

CHAPTER SIX

North Korea celebrated its first Day of the Star, a national holiday to honor the rise of Kim Yo-Jong to Most Supreme Eternal Leader of the hermit kingdom. Thousands of North Korean celebrants danced in the streets and, in an apparent deference to Kim Il-sung, showered the Kumsusan Palace of the Sun with flowers as done during Day of the Sun.

General Geyeon entered the royal dining room, head slightly bowed. His skin was an unusual shade of pale and he looked burdened by the large military coat covered with gold stars. The Most Supreme Eternal Leader was standing before a buffet of food spread out across the table. She moved slowly back and forth, picking at food here and there and occasionally placing a morsel in her mouth. She peppered the several chefs standing nearby with bursts of criticism, and after a particularly severe tongue-lashing, threw a plate at her head chef.

The Most Supreme Eternal Leader stopped mid-sentence when she noticed General Geyeon approach. Her pallid coloring was usually only contrasted by the bright red lipstick she favored. However, upon noticing General Geyeon's entrance, Yo-Jong's cheeks began to show blotches of red. The General had always been attracted by Kim Yo-Jong. Her diminutive size and warm smile were offset by a cold firmness of her eyes. He sometimes found himself turned on in her presence. He was unable to place whether it was her plainness which lent itself to a sort of innocence, or whether it was that look in her eyes that suggested an ability to shove a knife into the body of another, foe or friend, without second thought.

Now, with the increasing red blotchiness appearing on her face, she reminds me of a Rose of Sharon, he thought. *How much I would like to whisper that into her ear; to see if she would accept my compliment,* his thoughts paused for a moment. *Or slit my throat for comparing her to the National flower of South Korea.* He shuttered at the thought.

She watched intently as General Geyeon approached. Anger continued to rise in her face. The General avoided eye-contact and noticed a piece of yellow egg paper that had been hanging off the shirt of the head chef fell onto the shattered platter of gujeolpan strewn on the ground before him.

"What is this? You come before me without being summoned? Interrupting my meal?"

"Most Supreme Eternal Leader, it is only due to the gravity of the situation that I have disturbed your most great and precious time. It seem that we have—"

At that moment three American soldiers entered the chamber, followed closely by a half dozen North Korean military personnel. Kim Yo-Jong immediately assumed the horse stance, as if to combat the American intruders herself should they continue their advance. She turned to one of the several soldiers who had been stationed in the room, still standing at ease, as if no threat were present.

"Fool! Do not just stand there!" she screamed, turning her attention toward General Geyeon again. "How do you let these foreigners into the palace? Kill them!" She turned once again to her soldiers in the room, who had now begun advancing toward the group, weapons drawn. "Kill General Geyeon along with these enemy soldiers! Kill them all!" Her eyes were darting around the room and a hint of panic had crossed her countenance.

Col. Leeds and two Niklas continued walking towards Kim Yo-Jong. Col. Leeds held his hand up. "If you please, Most Supreme Eternal Leader. We are not going to harm you."

Kim Yo-Jong suddenly released her body from the horse stance she had been holding and directed her attention towards Leeds. The terror was no longer visible on her face. "Harm me?" She smiled. "Harm me?" A loud and raucous laugh suddenly emerged from her throat. "You think you can harm me? That is a joke." She pulled a North Korean made Type-70 handgun from her waistband and fired four shots at the Niklas to the right of Leeds. The bullets pierced through Niklas' torso, blood and tissue spattering out his back side. The Niklas stumbled briefly and placed his hand over the fresh wound. He wiped his hand away and as quickly as he moved his hand the bullet wounds healed themselves. The redness from his face quickly drained, leaving a healthy hue in its place. "Then it is true," she mumbled, arms falling limply to her sides.

Leeds stepped up to Kim Yo-Jong and removed the gun from her loose grip. "Please, Most Supreme Eternal Leader, do not force me to do something you would regret. I have told you we would do you no harm. And yet you failed to show us even the slightest courtesy. I will forgive you this once, but please do not test my patience again." Leeds placed the weapon on the table in front of Kim Yo-Jong.

Kim Yo-Jong flopped down into a chair. She started to speak, but no words came across her lips.

"Do not worry, Most Supreme Eternal Leader. I will do the speaking and you can do the listening." Leeds helped himself to a jeon and placed several selections from the delicacies onto the middle. He rolled the jeon and took a bite. "Delicious." He chewed slowly, savoring each bite as the head chef attempted to suppress the slight grin on his face. Kim Yo-Jong remained sitting, speechless, and General Geyeon held his ground, making no move either accommodating or threatening. Leeds finished the last bite.

"Now, Most Supreme Eternal Leader, what I have to say is quite simple. I have been tasked with opening lines of communication between our nations." Kim Yo-Jong's eyes opened wide and she leaned towards Leeds. Leeds held up one hand, and looked around the room. "We must be discreet."

"Be gone!" Kim Yo-Jong shouted, making a sweeping motion with her hand and the North Korean soldiers and chefs departed, except for General Geyeon who remained motionless. Leeds looked at him from the corner of his eye, but continued speaking to Kim Yo-Jong. "In exchange for your future help, I am willing to provide you with a single injection of PreventAll." The Niklas on the left stepped forward and removed a syringe from his chest pocket. He placed it into Leeds' waiting hand. "No one must know of this, of course." Leeds turned his head towards General Geyeon.

Kim Yo-Jong turned to the General. She picked up the gun lying on the table and rapidly fired two shots into the General. Geyeon staggered back briefly before collapsing to the floor, blood pooling beneath his body. "Go on," Kim Yo-Jong said.

"You are a smart woman, Most Supreme Eternal Leader. I am quite certain that you recognize that the only reason China has been providing you with assistance is due to their effort to keep you underfoot. The American government is fooling itself. I am well aware of the strength of your army and the advancements of your scientists. It is important for me that our countries remain outwardly hostile, but I am prepared to work behind the scenes, Most Supreme Eternal Leader, to ensure North Korean dominance in Asia and beyond."

Kim Yo-Jong nodded slowly. "For what purpose?"

"I am not willing to share that information with you at this time, but I assure you, if you work with me, the Democratic People's Republic of Korea will bring China to its knees."

Kim Yo-Jong smiled briefly. "Why should I trust you?" she questioned.

"You have no choice, Most Supreme Eternal Leader. There is nothing either you or your armies can do to harm me or my men, or to prevent me from accomplishing my goals. Frankly, I am not asking your permission, simply your cooperation, as my plans can be executed much easier if you are a willing partner, Most Supreme Eternal Leader. But, trust me, it can be done without your . . . shall I say, presence?"

Kim Yo-Jong's brow furrowed and her face flushed in anger. "How dare you speak to me like that!"

Both Niklases drew their weapons and Kim Yo-Jong looked around. The anger eased on her face as she remembered that she was alone, and that she could not do harm to the Americans anyway. Leeds raised his hand, offering the syringe. "It is your choice, Most Supreme Eternal Leader. Will it be immortality, or the alternative?" The Niklases each cocked their weapons.

Kim Yo-Jong nodded slowly.

"You will raise the Democratic People's Republic of Korea above all nations in Asia," Leeds said, depressing the PreventAll syringe into Kim Yo-Jong's arm.

She relaxed a moment, then tensed up as the anger returned in a sudden change. "I do not feel different!"

"And why should you?" Leeds smiled wryly.

"You have tricked me!" She leapt at Leeds. He raised his arm and struck her across the face as she neared. She landed on her feet in a crouched position and swung around and tripped Leeds with a leg sweep. Leeds fell to the floor and as Yo-Jong attempted to pounce upon him she was intercepted by a Niklas. The two crashed to the floor, Niklas landing on top. He pinned her shoulders to the ground and held them there with his knees. As she raised her legs to wrap them around Niklas' head, Leeds stood above her and pointed his gun at Yo-Jong's forehead.

"Feisty bitch, aren't you?" he said. He emptied his clip into Yo-Jong.

Yo-Jong felt an intense heat in her brain and smelled the scent of spent gunpowder. She could feel the blood rushing from both the front and back of her head, soaking her face and neck in an acrid smelling fluid.

And then nothing. There was silence, but rather than the expected darkness, she saw the image of Niklas, still astride her. "This is not what I imagined death to be like," she whispered.

"I imagine not," Leeds said, from above her.

She turned to look at Leeds, confused.

"So I ask again, why should you feel any different now that you have been inoculated?" Leeds asked, and then laughed.

Niklas had risen back to his feet and Yo-Jong, still lying on her back, reached up and felt her face. It was dry. She could feel no evidence of any injury. She leapt up and ran to where one of many decorative mirrors was inlaid upon the wall. "Amazing."

"Before you start imagining that you are now my equal, be aware that I have the ability to deactivate the nanobots that are giving you immortality." Leeds pulled out a small electronic device and pressed several buttons and then turned to the Niklas on his left. "I am sorry, Sergeant."

The other Niklas immediately drew his gun and pointed it toward the first Niklas and pulled the trigger four times rapidly. The Niklas let out a groan and collapsed to the floor, the blood pooling around him. He convulsed several times and then abruptly stopped moving. The second Niklas bent down and felt his carotid artery. He looked back to Leeds and nodded gently.

"Now, Most Supreme Eternal Leader," Leeds said, turning back towards Kim Yo-Jong. "As long as you cooperate you have nothing to fear."

CHAPTER SEVEN

The armies of the North Korean nation continue to build along the the Yalu River from Dandong, China. Emperor Xi has threatened to cut off aid to the hermit nation if aggressive action is not halted immediately. North Korea's dictator Kim Yo-Jong denies any nefarious motives and has stated that the build-up of troops is solely for the purposes of distribution of a North Korean developed vaccination against Covid-19-B2 which is being produced in its high-tech laboratory in Soryong-Dong. Chinese intelligence states it has no information relating to any such lab and has begun build-up of massive amounts of military forces in an attempt to display the overwhelming might of the Chinese military.

Shortly after her arrival in Israel, Eva quickly fell into a routine that enabled her keep her mind off the events of the first night. She was provided with an apartment, the free use of public transportation and full access to the laboratories at Tel Aviv University where teams of Israeli scientists were hard at work on a number of various projects of which Eva knew little about and cared even less. She was assigned to a small group of scientists who were cooperating with DARPA in the development of improvements to PreVentall. Eva was unaware if her fellow scientists understood how their research could be used to improve the communication experiences between the Niklases in an effort to advance the effectiveness and efficiency of the Niklas army, or whether they even knew it was an aspect of PreVentall itself. She found her Israeli colleagues to be a bright and affable group and, much to her surprise, found herself thinking of them as friends, rather than colleagues. Her surprise was not that she had grown personally fond of them; she found the Israelis to be witty, smart and surprisingly laid-back, unlike many of her American counterparts, which tended to be dullards and somewhat uptight when it came to complying with the many rules required by DARPA projects. Perhaps it was the fact that they had all served mandatory time in army and were used to the structure and the need to take most decisions up the chain of command, or, more likely, follow those decisions coming down the chain of command. Her surprise arose from the swiftness with which she had bonded to this small group and how quickly they adopted her as one of their own.

Eva had expected that the Israelis would be cooperative and, perhaps, even cordial. However, the world outside of North America was suffering a great disparity when compared to North America. While news coming to Americans from abroad was limited by what could gleaned from iMeme feeds, Eva took full advantage of the freedom of information existing outside the highly censored environment of the United States. It was clear that many of the benefits Genesis technology was to grant to the outside world (and which were highly touted when Genesis was first introduced to the American audience) never came to fruition. While American Genesis technology allowed for the free manufacture of products, many American businessmen dealing in foreign markets took advantage of the situation by not only lowering prices of the goods they sold, but by lowering wages offered to their foreign employees. Workers were confronted by a marketplace with the price of goods lowered by half and salaries lowered by an almost equal amount. The inequalities of capitalism continued to plague the outside world and its closest companion, war, appeared to be alive and well. And lurking in the shadows.

The sun had risen into a clear morning sky and along with it the temperature. Eva usually enjoyed her morning commute from her little apartment on Snir Street to the Tel Aviv University campus and this morning was no different. Her apartment was just about one mile away (*Two point two kilometers*, her Israeli counterparts would correct her). Like most Americans, she never could get used to using the metric system in everyday life. Eva smiled as she watched some children in the neighborhood running down the street, their backpacks bulging with who-knows-what, laughing and yelling. As she crossed the Ayalon Highway and approached the campus, the pedestrian traffic changed from families and young children to university students and professors. *Boy, it's unusually humid today*, she thought as she felt perspiration build up in the small of her back. She slowed her pace down considerably in an effort to avoid being covered in sweat by the time she reached the lab. *I don't care what they think*, she mused. *Who wants to sit in an air-conditioned car on their way to an air-conditioned lab anyway? Nothing is better than fresh air, even if it's a little sticky.*

When she finally arrived at the lab, her new friends did not disappoint.

"You're late," Rachael said, looking up from the computer screen she had been staring at. Rachael Sarfati was the lab group leader, although many of her tasks involved matters to which the other group members were not privy. She was twenty-seven years old and single. Her energy seemed to be bottomless but Eva suspected that her quick wit and good-natured humor hid a darker side Rachael would not let show. She kept a certain distance in her relationship with Eva and had an air about her that made Eva glad that they were friends and not enemies.

Due to the nature of her project, Eva also often worked independently, using the others only to perform work that advanced her research, but would not, independently, reveal the nature of her work or disclose any secrets unique to PreVentall.

"I guess I was walking a little slower today," Eva responded.

"Lazy American," Rafi chimed in. Rafi Goldner was the older brother Eva never had. He was protective of Eva and went out of his way to ensure that she felt a part of the group. But like an older sibling, good-naturedly seemed to like to give Eva a hard time.

"Don't be rude, Rafi," Lars spoke, matter-of-factly. Lars Daalman, a German of Dutch descent, was the outsider of the group. While he participated fully in the lab and was welcome at all their gatherings, Eva felt that he was never really treated as an equal in the group. He was the only non-Israeli, other than Eva. And although he was, no doubt, quite brilliant, he seemed to try too hard. He did little to endear himself to Eva, continuously talking about the Americans' brashness and letting slip his anger at their refusal to share the secrets of PreVentall. He also tended to speak a little too often about the inequity of Americans refusing to inoculate non-US citizens. Eva did not blame him for the desire to participate in the benefits of good health and eternal youth, but his manner of expressing his disappointment came across as whiny and petty.

Yitzak turned and smiled at Rafi. "Ignore him, Eva. Hebrew is an ancient language kept alive for many centuries in the bible only. With so many 'the Lord, our God' and 'God of Abraham, God of Isaac, God of Jacob,' all this repetition is built into our syntax. He did not mean to use the tautological *lazy American*, but we Hebrews cannot help but say everything twice. That's why we make such good lawyers. We learn from an early age how to double-speak; null and void, cease and desist, and all that repetitive nonsense." He laughed a hearty laugh. Yitzak ben-Aharon then turned towards Eva and gave her a wink. His looks and youthful physique defied his age and smoking habits. At thirty-six, he was the oldest in the group. He enjoyed the company of the group immensely, but often skipped group get-togethers to spend time with his wife and two children. He kept pictures of his wife Naomi and his two young children on his lab bench. Yitzak kept careful eye out for everyone in the group and often played the father role for their group, making sure everyone got along, or at least did not hurt each other. He was inherently kind and Eva sometimes found herself staring at him. *I wonder if I'll ever find someone like that for myself. Naomi and their kids are lucky to have him around.*

Eva found the Israelis generally to be a puzzling bunch. They all seemed to smoke and eat foods not conducive to a healthy lifestyle, yet none seemed to suffer any ill-health affects as a result. They were, by necessity, an active people; serving in the army and seemingly all enjoyed hiking, walking or other physical activity. Perhaps this explained the seemingly higher level of health and lower level of obesity than she recalled in American society prior to the introduction of PreVentall. She also credited the Israeli government for determining that basic health care was a fundamental right. As such, Israel provided its citizens with universal and mandatory coverage. In addition, every year the Knesset would review those medical conditions which were not included in the purview of basic coverage to determine which heretofore uncovered conditions would be added, and the budget appropriately amended to cover the increased outlays. This system resulted in high satisfaction among Israeli citizens and a deep confidence in the medical system as a whole. It also resulted in efficiencies that were credited with higher health outcomes than other first-world nations.

Eva witnessed this first hand. With the threat of the Covid-19-B2 virus becoming a serious epidemic within Israel's borders, the government immediately commissioned scientists to develop a vaccine to protect its citizens. Once discovered, massive resources were dedicated to produce sufficient vaccine to inoculate the entire population.

It had not taken Eva long to learn what the ubiquitous posters with the red spot with the man's face were. The red dot was the Covid virus and what she took to be representative of a man's face was actually the Hebrew letter Bet. These were notices of the mandatory Covid-19-B2 virus vaccines required of all Israeli citizens and visitors, except, of course, the few Americans entering the country - such precautions being made unnecessary by PreVentall.

Among the few non-Israelis with which Eva encountered in her research, and with Lars in particular, there were always whispers that the Israelis had been somehow involved in the development of molecular manufacturing and some of the technologies behind PreVentall. Rumors abounded that the Israelis, in fact, had some incarnation of the PreVentall.

"I have heard that the Israelis are all inoculated," Lars confided in Eva one evening while they were both working late in the lab.

"What are you talking about, Lars?"

"This ruse about the Covid-19-B2 virus vaccine. Haven't you noticed how except for individuals entering the country from abroad, that the country is being immunized in sectors? That does not make sense from a pandemic standpoint. There has been no shelter-in-place orders and people are traveling between sectors, yet we don't hear of many Covid-19 cases happening here in Israel. Why is that?"

Eva paused for a moment. She had not considered the prospect. "I imagine that within the sectors, immunizations are taking place on a priority basis, protecting the most vulnerable first, then moving on to the healthy populations."

"Yet *no one* has been infected? Look at Jordan and Syria. It's wiping out entire villages," Lars responded.

"Herd immunity, Lars. Once a large percentage of people have the antibody, the virus cannot find enough hosts to continue to spread and even the un-immunized don't acquire the virus. They have been immunizing the population for months. It works. Neither you nor I have caught the virus." She hoped that would end the conversation.

"Of course you haven't, Eva. You're immune. But I haven't because," he paused. "Because everyone we work with and that I'm around are inoculated. It's like herd immunity, except it's PreVentall immunity." Lars was visibly upset.

"Lars. If the Israeli's were immunized, they wouldn't age. They wouldn't get hurt. Remember last week when Yitzak brought Ben to the lab? He fell off a chair and bruised his head pretty badly. That's not what happens with PreVentall."

"It's different, here. Don't you see, Eva? How can you not recognize it? They are getting inoculated. Look at how many Israelis smoke? And the foods they eat? My goodness, it makes Speke look healthy in comparison. Why--"

"Speke?" Eva interrupted.

"You know, lardo?" Eva continued to stare at Lars with a blank expression. Lars' anger softened. "Think of bacon, except without any meat, just the fat." Eva grimaced. "Ahh, my dear. The Germans love it. Almost as much as Griebenschmalz."

Eva raised her eyebrows. "Now you're making stuff up."

Lars laughed softly. "I must get you to Germany someday. But never mind, you take me off track. The Israelis must be inoculated. They--"

"Hold up, Lars," Eva interrupted again. "You and I both know you can't step 100 feet outside of this lab and not run into a half-a-dozen professors clearly in their sixties. And this county, including most government officials, is filled with people who are clearly not inoculated."

"No, no Eva. You misunderstand. It is not the same PreVentall you Americans enjoy. It is different. One that does not prevent aging, but does keep one healthy. Perhaps stronger, faster, more able to defeat their enemy. Perhaps only giving the appearance of normalcy. For some type of military advantage."

"I don't see it, Lars," Eva responded.

"Yes, it is difficult to grasp. But have you noticed anything peculiar about every Israeli you have met? Certainly, as an American, you should be much more familiar than I, who has merely taken notice from having kept company with several of the U.S. Military personnel who, at least here at Tel Aviv University, seem to outnumber that of the Israeli Defense Forces - which I also find quite telling."

Eva remained silent.

"The silence," Lars said, smiling with the confidence of a magician who is explaining the secret behind his greatest illusion.

"What are you talking about, Lars?" Eva questioned, decidedly confused.

"When you meet an Israeli, when you look into their eyes, do you not see it? When you speak to them, do you not hear it?" Lars' smile remained affixed to his face, his eyes reflecting the hope of understanding.

Eva thought for a moment. She could recall no characteristic, spoken or expressed, that would set the Israelis apart from any other non-American. "Put a thirty year old Israeli side-by-side with an European and I couldn't tell you which was which," she stated.

Lars let out a small laugh. "Perhaps you Americans don't see it. Have you ever been asked by an Israeli for PreVentall?" Eva shook her head, as Lars continued. "No, I'm certain you have not. And have you noticed how they have no reservations in acting as an equal among Americans?" Lars' eyes seemed to beg understanding.

Eva thought back to the many times Lars defended her, went out of his way to help her out. It suddenly struck Eva that Lars sought some form of sympathy and communion. She had long considered that his efforts were an attempt to gain her friendship, and perhaps her intimacy. Looking even now into Lars' eyes she suddenly recognized that they did not shine with the soft glow kindness, but smoldered with the flames of desire and jealousy. He wanted PreVentall.

Lars continued. "I have heard that the Israelis are all inoculated with some special formulation of PreVentall. It has long been suspected in many parts of the world that the Israelis were very involved in the development of certain aspects of the Genesis project. It would not be past them to co-opt some of the technology for their own purposes. It would be quite a military advantage for them, and, perhaps beneficial for Americans." Lars paused a moment. "You know, it would not hurt for the Americans to share the technology with some of its European allies as well." A weak smile passed his lips.

Eva made a mental note to be more careful around Lars, as she slipped her hand into her pocket. She thought back to her conversation with Col. Leeds and his advice as he handed her the vial of Forget-Me-Nots. She made a second mental note to start keeping a vial with her at all times.

"Perhaps there are reasons the Israelis would be included and the Europeans not." She had meant it as a commentary on the current rising hostility against Israel from Iran, but she immediately recognized that Lars was too serious about the belief of Israeli access and she regretted the comment as soon as it had finished emanating from her lips. "I mean, perhaps, *any* nation that is under constant threat of annihilation has more use of PreVentall than those who are not." Lars seemed no longer to be listening, the fire of jealously raging in his light eyes.

"No one outside of North America is getting inoculated, Lars. And even if they were, I certainly don't have the authority or ability to decide. So just let it go. L'Hitraot," Eva said and turned to go. She stopped and faced Lars again. "You know, they still use that. L'Hitraot. That doesn't sound like PreVentall talking to me." This time she turned and did walk away.

The tension that had existed between Israel and its Arab neighbors since Israel's inception was an ever-present cloud that hung over the nation. But the Israeli people did not appear to let it consume their day-to-day lives. They found ways to adapt to the reality and carry on with the semblance of normalcy. Eva had learned quickly that the Israelis did not say goodbye to each other when parting, but rather they would say, *L'Hitraot*, a phrase meaning *See you later*. This simple phrase conveyed not only the hopes and prayers that no harm would come to either the speaker or the recipient, but also avoided the stark reality of the phrase *Good bye* – words flush with the very real possibility that it could, in fact, be the final meeting between family or friends. Eva sympathized with the harsh and cold certainty such a phrase could hold for any person who lived under the constant shadow of annihilation.

Eva's time in Israel was productive. She worked tirelessly as she had always done, putting her full concentration into the matter at hand. She looked at each task she was given as a personal challenge and felt a sense of pride with each new discovery she made, although she never felt quite the same sense of innocent satisfaction she felt after helping solve the problem with the California Condors. Since she had been recruited by DARPA her tasks all involved matters that reached far beyond her imagination, tasks that inevitably held a darker side. *Duel purpose*, scientists liked to refer to such matters as. Her involvement in the California Condor Project morphed, unknowingly to her, into a project to duplicate humans. The project's success, as well as its darker underpinnings, was evident by the army of Niklases that now staffed a significant portion of the American military forces.

The problem she was now tasked with was attempting to devise a method in which multiple beings could share a single consciousness. This was, as Leeds had pointed out, more than simply a question of communication. While Eva was not familiar with any studies relating to the subject (if there even had been any), she was certain that after replication, each Niklas continued to make new neural connections that would not only prevent other Niklases from sharing identical thoughts and ideas after more than a few days, but could alter each individual Niklas' personality. Regardless of where one came down on the subject, it could not be denied that both nature and nurture have great influence on the person an individual becomes and into whom they continue to evolve. But Leeds' seeming secrecy around this whole matter raised concerns with Eva. She was no military genius, but she could not help but wonder where the implications of this new challenge would lead. Knowing Col. Leeds, it was certainly larger than what he had led her to believe. But she managed to keep her concentration to the task at hand, and her mind wondered into the implications only on those evenings when she was not so exhausted that she fell asleep as soon as her head hit the pillow.

CHAPTER EIGHT

The Jerusalem Post: Reports from across the region reveal that the current Covid-19-B2 outbreak is causing devastation to the Jordanian and Lebanese populations. Jordan's death toll has climbed to nearly 250,000 while the death toll in Lebanon has reached almost 200,000. Global infection rates are on the rise and the United States is sending troops to the region and across the globe in an effort to combat the rising risk to human population.

Benjamin M. Safrad Day School is housed in an innocuous building on the corner of Hassan Shuqri and Berwald Street. Except during inclement weather, there was usually a group of young children outside on the playground. While most of the children are dropped off by their parents on the way to work and picked up in the evening, Safrad was originally created as an orphanage. Inevitably, the several wars Israel found itself in always resulted in the unfortunate creation of orphans. Although Israel found itself frequently in the throws of a long-running cold war with its neighbors, it had been several years since it was involved in a hot war. This reduced the need for orphanages and Safrad had enjoyed the benefits of providing education for children with families to return to at night. Miriam Gila, the school's chief administrator, was looking out her window watching the many parents picking up their children for the evening. As she watched the last of the children leaving she was startled by the sound of a knock on her office door. She turned around to see two uniformed soldiers standing in the reception area.

Her secretary was standing in the doorway. She wore a slight frown. "Miriam, there are a couple of gentlemen here to see you."

"Thank you, Adele," Miriam said, directing her gaze to Adele. Adele avoided eye contact as she stepped back out of the threshold to allow the two soldiers entrance.

"Madame Gila?" the soldier on the left asked.

"Yes, how may I help you?"she replied, stiffening.

The soldier on the right took a step toward Miriam Gila and held out his hand. "I'm Major Arbell. I'm leading the Covid-19-B2 immunization program for Area 3, which includes your school."

Miriam let out a sigh and noticeably relaxed. "Major, thank you. I was afraid that you were bringing me an orphan." She clapped her hands together and a smile appeared on her face. "It has been some time and, well. It's a terrible situation. I'm sure you understand. But Covid-19-B2 vaccines, that is good news. We seemed to be near the bottom of your distribution list." She gave Major Arbell a slight frown.

A forced smile crossed Major Arbell's face. "Yes, but children are in a low probability group. We have arranged to have medical staff visit this facility tomorrow to inoculate all the children, as well as any adult staff who have not yet received their immunization." He stood silently for a moment. "However . . ." He looked towards his fellow soldier, causing Miriam to look too.

Her face dropped but she quickly lifted her head and put on a large smile. "And what is your name?" she asked stooping over and reaching out her hand. Miriam had initially failed to notice the young child that was standing, mostly hidden, behind the second soldier.

The little girl stared at Miriam but said nothing. She tried to move back behind the soldier but he held her hand, firmly but gently, and prevented her retreat.

"I'm Ms. Gila," Miriam said, crouching down to lower herself to the child's level, her arm still outstretched.

The little girl looked up at the soldier, and then began to reach her hand out towards Miriam, but seemingly changed her mind and raised it to her lip. She looked down and spoke quietly, "Noa."

"Well, very nice to meet you Noa." Miriam pulled her hand back and patted smooth her skirt as she stood again to continue her conversation with Major Arbell.

As she did so, a young woman entered the room carrying a small pink suitcase. She took Noa with her free the hand and led her towards the staircase leading to the second floor living area.

"My name is Sarah. I'm going to show you where your new room is." They walked up a staircase and entered a room with two beds, two dressers and an open shelf cubby with several worn board game boxes and a few toys scattered among the many empty spaces. Sarah released Noa's hand and placed her suitcase upon the nearest bed. Noa walked tentatively around the room, exploring the space carefully. After satisfying her curiosity, she squatted down on her haunches and crossed her arms over her knees. She placed her chin on her arms and watched as Sarah removed the several articles of clothing that Noa owned, breathing in deeply the slight scent of almonds from some lotion or perfume Sarah wore. Noa fascinated herself with the gentle movements of Sarah as she refolded clothing and placed them in drawers.

A rolled up pair of socks dropped from Sarah's hand and moved slowly across the floor boards. Noa followed its path as it made its way away from where she was squatting. That is when she spotted a pink stuffed rabbit beneath the bed. She immediately crawled over to retrieve the rabbit, holding it tightly in her arms.

"Oh, bunny. Are you all alone like me? I will hold you and protect you." She squeezed it tight and closed her eyes, twisting side to side.

Sarah bent down. "You're not alone, Noa. I will be here with you. See that door over there?" She pointed to the side of the room toward a door that was ajar. Beyond the threshold Noa saw the edge of a bed and a dresser with a mirror. "That's where I stay. And if you need anything at all, I will be nearby."

Noa nodded at her. "I think I will like that."

The building was mostly empty when Sarah and Noa returned to the main floor. A custodian was emptying trash cans but the place was otherwise quiet.

"Are you hungry?" Sarah asked Noa. The little girl nodded. They walked into a large room that was mostly empty. Several tables with attached benches were folded and lined up against the far wall. Several large trash cans stood in a corner. One of the walls had a couple large bulletin boards on which were various colorful sheets of paper announcing news of one sort or another. On the opposite end, the room was divided by a counter with a kitchen on the far side.

"Well, I really wasn't expecting you so it looks like we'll have to make do with what we have." Sarah began opening cabinets to see what options they would have for dinner. She spotted the familiar blue box with golden macaroni noodles on a fork. "How about a Kraft Dinner?" she asked, her eyes growing large. Sarah knew few children who didn't love macaroni and cheese. This elicited the first smile Sarah had seen from Noa who was nodding her head excitedly.

Dinner had been quietly shared between Sarah and Noa, who occasionally offered a bit of her macaroni and cheese dinner to her pink rabbit friend. After dinner Sarah bathed Noa and they watched a little television before turning in for the night. Sarah put on her pajamas too, and after tucking Noa in, slipped into her bed to read a book and to wait. The first night at Safrad was always the hardest on new children and Sarah knew she needed to be nearby, not if, but when Noa would need comfort.

* * *

The morning started quietly enough but fell into chaos for Miriam. A suicide bomber had threatened to detonate an explosive device on the Egged Route 245 bus heading toward Kiryat Rabin and traffic in the area was at a standstill. It was already past 9 o'clock in the morning and Miraim was three cars behind the bus and was surrounded by military and rescue personnel.

"I know the Covid-19-B2 inoculations are today," she said, speaking to her iMeme and looking around to see if there was any opportunity to move from her current position.

"The technicians have already started the immunizations," the voice on the other end continued. "Millie Rosen called in sick this morning, but otherwise the other children are all accounted for."

"Well, I think I'm going to be stuck here for some time Sophie," Miriam sighed. "Just please ask whomever is in charge if I can schedule an inoculation with my personal physician. We will need to figure out how to handle Mille, too, if she has not already been inoculated."

"Ok. I will." the voice replied. "And, I know you're stuck out there in that mess, but thank God that terrorist failed. A little traffic is a good trade-off. I'll see you later, Miriam."

The phone line disconnected. *She's right,* Miriam thought.

Sophie hung up the phone and turned back towards her own chaos unfolding in the lunch room. "Okay children. After you receive your inoculation, you can stand over by the playground door. Those still waiting, please try to stay in line. Give the nurse your name, please."An ever-growing group of children stood by the door and restlessly were moving about. The line of children awaiting their Covid-19-B2 shot were even more restless. Pushing and shoving was constant and several children would leave the line when they saw friends from other classrooms. "Children! Enough! Please stay in line." It was as if she had not even spoken. She turned toward a woman standing near the playground door. "Jennifer, let's just let those who are done go outside. You and the others can go out there with them. I think I can handle these kids inside." Jennifer gave her a doubting look. "Okay, well . . . it can't be much worse." They both laughed.

Noa, standing toward the back of the line, was not noticed by the other children. She had hoped to see Sarah but she had not returned to the school. Noa was too young to realize that Sarah had a real home and a real bedroom and that her room at the school was part of her job. It would always be a few weeks until the new children became familiar with the fact that Sarah was only at the school from 5 p.m. until she left at about 6 in the morning, when the regular staff started filing in. Seeing the other children playing outside, Noa managed to work her way slowly from the group of children awaiting their Covid-19-B2 vaccination, though the kitchen area she and Sarah had been in the night before and to far side, near to the group of children who had already received the vaccination. It was when Jennifer opened the door, her attention captured by the children pouring outside that Noa slipped unnoticed from behind the counter. Once the doors were open and kids started filing out, she joined the group outside and made her way towards the edge of the playground, clutching her pink rabbit. Jennifer was walking around checking on the children when she noticed Noa. *Who is that?* she thought. She did a quick check of the rest of the playground, but did not notice anything out of place. The other teachers were watching their respective areas. Jennifer began to walk towards the new child, who was now involved in a conversation with two other girls. As she neared the trio she heard the sudden sounds of children screaming and turned to see Michael and Yoni in a fight. She did not see any immediate danger in leaving the mystery of the unknown child for a moment and directed her attention to the scuffling children. As she pulled them apart she saw that Yoni was bleeding from a cut above his eye and she scuttled him back into the building. The personnel were just finishing up the inoculations of the last few children. "I apologize, but can one of you please take a look a Yoni's eye?" One of the medical techs stood up and took the child from Jennifer to administer first aid as Miriam entered the room.

"Miriam! Just in time to miss the chaos," Sophie laughed. Miriam rolled her eyes.

"Miriam," Jennifer said. "There's a young girl out there," she turned to point in the direction of Noa. "I—"

Miriam turned to look. "Oh, that's Noa. She is a new foster. She arrived last night. I apologize, I would have thought Sarah would have introduced you before she left."

"We must have missed each other in the confusion. The Covid-19-B2 folks arrived early and we started running around to gather the children and, well, it's been hectic ever since."

"Well, she seems to be making friends already. I think that's a good sign." Miriam stated.

"I agree. Hopefully she's on to better things," Jennifer said, and moved back to the doors to return to the children outside.

CHAPTER NINE

The Israeli government announced today that it expects completion of the Covid-19-B2 inoculation program to be finished within the next several days. Hostilities remain high among the Israeli government and the Arab population and the army has made multiple reports of Arab population refusal to cooperate in the program. The IDF has agreed to allow the Palestinian population to cross the border with Egypt and Jordan to receive the Covid-19-B2 immunization from Arab physicians. Reports are that a majority of Palestinians are taking advantage of this privilege, in many instances using the objection to Jewish physicians simply as an excuse to visit family living outside Gaza, Samaria and Judea

"Are you ready for the meeting today?" Rafi asked.

"Oh shit! I forgot about the symposium. Is that today?" Eva asked.

"You know it, Professor," Yitzak said. "You're prepared, right? They changed the format. You're giving the keynote address, remember?"

Eva raised her head quickly as she felt a sinking feeling weight in her stomach. "What!? When did they do that? I wasn't even supposed to present! Keynote address?" Eva stood up from where she was sitting, a mixture of terror and anger crossing her face.

"Of course you were. You asked for the results last week on Assay 7.3 and had my group address several concerns with the Gimel 9 interface," Rachael said.

"The data from Assay 7.3 hasn't even been analyzed," Eva said, her voice clearly struggling to remain steady.

"Of course it has. And here's the data from the Gimel 9 interface," Rachael responded as she set a pile of documents in front of Eva. "Right here."

Lars moved to say something, but stopped. Eva was flipping through the pile of papers and did not see Yitzak scowl at him as he motioned a threat of serious and immediate harm to Lars should he speak out.

"For Christ sake. When did we get this?" Eva asked, sitting back down and hurrying to absorb the information before her.

"It's been over a week. Are you telling us that you're not prepared?" Yitzak asked. "The great Professor Eva Diaz isn't prepared? Shit! There goes my career, along with hers," he said sullenly. "My mother always told me not to hitch my wagon to someone else's star."

"Don't fuck with us, Eva. This is important," Rachael said, perturbed.

"I . . . . I'm not kidding. I swear I never knew I was presenting today." Eva's voice was soft as she looked towards Rachael. Her eyes told the group she was not kidding.

"Oy vey iz mir! Ha-matzav - khara," Rachael responded.

Eva looked to Rafi. "She said, this situation is shit," Rafi responded curtly and then quickly looked away.

"Eva. I'm going to ask you something and I want you to respond honestly," Rachael said, controlling her emotions. "Are you telling me you aren't prepared for the symposium?"

Eva shook her head, which suddenly began to ache. She placed her face in her palms and tried hard to think. Her mind raced to remember some trace of conversation, some snippet of email, anything, relating to the presentation.

"Maybe just play like you don't know anything about it," Rafi said.

"Do you think that would work?" Yitzak asked.

"I think it might," Rachael answered.

Eva lifted her head out of her hands. "What? You think that I just act like I don't know anything about it and you think that's going to work?"

"Who is playing?" Rafi questioned. "Are you playing, Yitzak?"

"I am," Yitzak said. "Are you, Professor Sarfati?"

"Me? Playing what?" she responded.

"Aren't we playing?" Rafi asked. "Or are we done playing?" The corners of his lips began to curl.

"No, we are playing," Yitzak insisted.

"Playing with her mind," Lars said. Eva looked quickly at him. Rafi's face turned serious again and he punched Lars.

"Ouch!" Lars rubbed his shoulder where it was hit.

"You're a shit, Lars. We had her going," Yitzak said.

"Oh, did we! She was gone," Rafi said as he began to laugh.

"You should have seen your face, Eva," Rachael added as she and Yitzak also broke into laughter.

"You assholes!" Eva shouted as the adrenaline response began to subside and her heartbeat began to return to normal. She let out a sigh.

"Oh, it was all in fun." Yitzak smiled, trying to calm his laughter and Eva's anger.

"A real ball," Eva said unconvincingly, pushing the stack of papers aside.

"Hold on there, those really are the results of Assay 7.3. We just finished them this morning," Rachael said, moving the pile back towards Eva.

Eva started to thumb through the pages.

"Don't get too engrossed, Eva," Yitzak interrupted. "Professor Echud's group is going to present soon on the effects of quantum entanglement on neural cells in the replication process. He believes that replicated neural cells can retain parity through quantum mechanics. He's almost as big a loser as Lars."

Lars frowned. "I am not a loser."

"Tell that to Echud's group. We've got about an hour until the presentation, Eva" Yitzak said.

Eva stood up. She grabbed the results from the Assay 7.3 from the top of her desk. "I've got to stretch my legs." She gave the group a half smile and shook her head as she left the lab.

Eva meandered around the campus briefly. The day had gotten considerably hotter and she stepped into a nearby building. It was the Bar-Shira Auditorium, the location of the upcoming presentation. As she passed by the door to the auditorium stage, she noticed a small note card taped above the room number nameplate.

REAL-TIME BIO-INFO UPDATES AND APPLICATIONS TO HEALTH OUTCOMES. She had been spending almost all of her time contemplating the problem of parallel brain development. It struck her that if the Niklases' brains could be replicated, on a continuous basis, in a central computing lab, then if it were possible to take all the updated information and translate it back to each individual Niklases' brain, the end result would be that each Niklas would, in theory, share the exact neural connections as the others. The same experiences, the same thoughts, the same synaptic changes. But in order to allow the Niklases to function, and not be in a state of continual brain death and rebirth, so to speak, the process had to be limited to only those neural transmitters that actually underwent change. Further, Eva felt that these neural upgrades needed to be collected, processed and re-transmitted in such a way as to prevent organic changes that are the result of each Niklas' daily interactions from interfering with the updating of his neural transmitters from information received from other Niklases. It was still almost a forty-five minutes until the symposium started, but she decided to go in. She could use the time to review the Assay 7.3 data in private.

There was a small group of military personnel quietly talking at the foot of the stage, some holding a green folder in one hand. She recognized Col. Leeds. The others all appeared to be IDF and from the look of their uniforms, Eva guessed they were part of the top brass. They had not noticed her as she walked in. She placed her tall stack of papers dealing with Assay 7.3 on the edge of the closest table of a line of six tables which, end-to-end, stretched across the entire stage.

"Hey, Professor!" Lt. Cobbs shouted from the top end of the room. He was walking down the aisle, a large stack of green folders in one arm, placing them on the long lecture hall tables in front of each chair. The two rows of seats behind him had been completed and it was clear he would be doing the same for the remainder of the room. Col. Leeds looked up towards Lt. Cobbs, his face glaring, and then over his shoulder to Eva. The men had stopped talking and were all staring at Eva. Col. Leeds turned his attention back to the group briefly and they quickly disbanded. One of the IDF officers stepped towards Eva and she noticed his military jacket draped across the back of the chair closest to her. She moved quickly to grab his jacket and hand it to him, knocking over the large stack of Assay 7.3 papers, which proceeded to fan out across one end of the table. Eva did not know whether to collect her papers or remain there holding the jacket. She stared briefly at the mess before her, several of the pages now having fallen off the table and floated to the floor beneath. She blushed slightly and stepped down off the stage to meet the officer and hand him his coat.

"I apologize. I didn't think anyone would be in here and I was just looking for a quiet place to review my research," she said as she moved to hand the jacket to the elder officer. He was not a big man, but carried what were clearly a few extra pounds around his mid-section. She quickly drew the jacket back and instead held it open by the collar to assist the officer in getting the jacket back on. He turned and placed his arms through the sleeves and offered a quiet "Thank you" and a smile as he straightened out the collar and walked back in the direction of his fellow officers. Then men all walked out the opposite stage door, Col. Leeds exiting last and giving a quick look around the room before shutting the door behind him.

Lt. Cobbs had continued setting out folders in the row in which he was traversing and when he finished he set his stack of folders down at the end of the next row and proceed toward Eva. When he reached the area in front of the stage he looked around to make sure they were the only ones in the room. "You look and smell great, Doc. How about you and me skip this meeting and find an empty lab somewhere?"

"Oh, but don't you have more folders to pass out?" she said, feigning interest.

"I can finish later. Besides, I've had this terrible headache the past couple of days and I think, perhaps, you could help me cure it. You know, you still owe me one." He smiled smugly.

Eva's stomach turned at the closeness of his body to hers. She put on an air of concern. "I'm sorry to hear that you're not feeling well, Lt. Cobbs. Perhaps I can recommend a thorough review of your latest PreventAll update - you are Version 3.5, I presume?" She smiled at him as she turned toward the door and motioned him to follow. She paused to turn back and give him her best sultry look.

"That, Doc, would suit me just fine." He smiled at her as he leaned in close. "But, I have to let you know, doc, it's not the head on my shoulders that aches so badly."

"Well, then, we must make sure we take care of that right away." She grabbed his hand into hers and led him out of the room, being sure to let it go it when they stepped out the door. "We wouldn't want anyone talking, would we?" she smiled seductively. "Please, follow me." She led him out the doors. "We won't find any privacy in there, with the symposium starting soon. She led him to a nearby building and then down a hall before pausing in front of a closed door. She held her hand up as she peeked inside. Lt. Cobb stood still, his face still lit up by his smile. The lab was dark and empty.

"This should do," she said. She looked into his eyes as she led him in and quietly shut the door behind them. The motion detector triggered the lights, which flickered on but shed only a soft glow as the warmed up. Eva quickly flipped a switch on the wall and the lights went out. A frosted window on a door on the opposite side of the room let in a soft glow from the room next door.

"Hmmm," she moaned. She grabbed his hand to lead him across the room but he resisted. She could sense the excitement from Lt. Cobb. He stopped and pulled her close. She pushed against him, "No, not here. Over there, under the soft light." She motioned to the door on the opposite side of the room. "So I can see better." He let Eva lead him. She stopped in front of the door with the frosted window and turned to him. "I certainly don't want anything to be, uh, bothering you, Lieutenant. I'm going to give you all the relief you need." She moved her head close to his and smiled. She closed her eyes and lifted her chin slowly. As she did so, she quietly turned the door knob. As Lt. Cobbs leaned in towards her, eyes closed and lips parted she swung the door open and quickly stepped into the next room pulling him along. Lt. Cobbs nearly tripped over his own feet as he stumbled into the adjoining lab. He just caught his balance and the three scientists sitting among various lab equipment suddenly turn toward them.

"Lt. Cobbs has indicated that he is having some difficulties with--" she turned to him, "What did you say your problem was?" Eva asked Cobbs. He stood there, silent. His face turning slightly red in a mixture of embarrassment and anger. She quickly looked back towards Lars, Yitzak and Rafi.

"Oh, yes. I believe he indicated that he was suffering from headaches. I recommend a complete reinitialization of Version 3.5 along with a Lab 3 diagnostics to determine the possible causes of failure. Of course, if the Lab 3's come up negative, I would then suggest a Series VI protocol, just to confirm." She turned back to Lt. Cobbs. "I do hope the Lab 3's disclose any possible issues – the Series VI take almost 48 hours to complete and I'd hate to see you out-of-service for so long." She gave him a knowing smile. She turned to leave the room, but stopped just short of the doorway. "Until you diagnose, be sure to put Lt. Cobbs on quarantine. We can't risk anyone else if these are a radio-frequency errors that can be spread to other systems."

Yitzak and Rafi both stood up and moved to Lt. Cobbs' side. He made a move towards Eva, but was instantly put to the ground by Rafi. Cobbs had forgotten, albeit briefly, that Rafi was well-versed in Krav Maga and could inflict unmentionable harm if needed. Cobbs clenched his teeth and scowled at Eva. "You owe me bitch!" Rafi shifted his weight and Cobbs felt a sharp pain race up his arm to compliment the dull ache that the hold inflicted.

She knew he would be livid when he was finally released from quarantine in two or three days. The unlikely hope that she would be able to avoid him in the future flashed in her mind. *At least he finally got the hint that I'm not interested in him,* she thought. *Or at least I hope so.*

Eva returned quickly back to the auditorium to retrieve her Assay 7.3 research. As she began to reorganize the documents she noticed a green folder beneath the toppled stack. She set it atop her stack with the intent to return it to the pile Lt. Cobbs had left behind. As she stepped toward his abandoned pile of green folders, and moved to place the folder on the stack she noticed that the file she had was missing the black lettering on the others advising the future attendees of the name of the conference set to begin shortly. She flipped open the folder and saw it contained a small stack of papers. The top-most document had a heading that read: DARPA. Below, it was marked TOP SECRET: OPERATION PHOENIX. She looked up, startled. The room was still empty. She flipped past the first page. It was an outline with scattered notes scribbled along the margin.

TOP SECRET
OPERATION PHOENIX

- *Approved by PM and Key cabinet posts* ⇒ נכחיש במקרה של
- *MKs not cleared for knowledge* כישלון
- *Joint task exercise to take place as predetermined.*
- *Pretextual conflict underway*
- *Operation Disbelief at Stage IV ⇒ בעיצומה, כל המערכות עובדות*
- *Estimated immediate casualties > 2.7M*
- *Estimated long-term casualties < 0.01%⇒ קָבִיל*
- *Technion and T-A.U. scheduled operational on or before Target Date 2. Satellite Relay fully operational.*
- *Media Outlets to be –*

A chill swept up her spine. She stood over the stack of folders on the table when she heard a noise. She quickly grabbed several and slipped the folder she was holding under the top folder of the stack. She was standing there when the door on the opposite side of the auditorium made a loud click and swung open.

"Eva, what are you still doing here?" Leeds questioned, as he looked around the empty room. "And where is Lt. Cobbs?"

Eva looked up and tried to not look so surprised. "Colonel Leeds. Hello. I'm sorry, I thought you were done in here." She stood up.

Leeds continued to look around the room as he walked towards her. "You still haven't finished picking up your papers, I see," he said coldly, staring at the stack of folders in Eva's hands.

"Oh, yes," she uttered. She collected her thoughts and continued. "It seems Lt. Cobbs may have been having some difficulties with Version 3.5. I took him over to the lab to have it checked out. I just got back here myself, and thought I'd better help pass these folders out since Lt. Cobbs is unavailable." she said, hoping to sound much less nervous than she felt.

Leeds looked back toward the rows of seats rising above him in the auditorium. "He must certainly have had some issues, since I see he failed to complete his task. But that is not for you to complete. I will have a Niklas finish the job," he said as he looked suspiciously at the folders in Eva's hands. She made sure to tip them slightly so that he would see the face of the top folder and the clearly written black ink stating: REAL-TIME BIO-INFO UPDATES AND APPLICATIONS TO HEALTH OUTCOMES. Leeds' hard glare towards Eva relaxed as he continued to scan the room. "Let's get your papers cleaned up," he said as he crouched down to pick up the loose sheets of paper lying on the floor.

"Oh, no, Colonel, please. Let me do that," Eva said as she hurried toward the strewn stack of papers still holding the multiple folders. She slid them under the pile of of papers and straightened the stack up as Leeds stood and placed the several sheets he recovered on top of the pile. He began to speak when he was interrupted by a Niklas entering the room.

"Sir!" he bellowed to Leeds.

Leeds turned his attention to the Niklas. "Sergeant. Would you please continue passing out those folders to each seat in here? It seems Lt. Cobbs was unable to finish the job," Leeds said coldly.

"Yes, Sir." the Niklas said, then paused, as if thinking. "Sir, Lt. Cobbs has been quarantined as a precaution relating to Version 3.5. He is under our watch until test results are completed." He then proceeded up the stairs to continue distributing the folders.

Leeds expression seemed to soften somewhat. "If you would like, Professor, I can arrange to have a Niklas assist you in carrying that stack of papers."

"Thank you Colonel, but I'm fine. I need to review some of these results before the conference starts."

"I will see you then." Leeds turned and left through the door from which he entered, his eyes still scanning the room.

Eva let out a soft sigh. She was always uneasy being left alone with a Niklas and today was no exception. She watched the folders being laid out, meticulously, across the room. *Look at him,* she thought. *It.* She turned back toward the exit. *Please, don't let Leeds suspect anything,* she prayed to no particular god.

Eva carried the Assay 7.3 papers up the stairs and took a seat at the end of the last row in the left corner of the room. From that vantage, she was able to see the side door where Col. Leeds last departed. She set the stack of papers, with the folders beneath, on the floor next to her feet and took the top inch or so of the assay results to review while she sat.

The room began to come alive with chatter as various scientists and researchers started to trickle in, generally congregating near the seats at the front of the auditorium. Eva was left alone to her assay. As the time approached for the presentation to begin, one of Professor Echud's staff stepped onto the stage and gently tapped the microphone to check the audio volume. The audience began to quiet down and turn their attention towards the stage.

Eva turned her gaze toward the stage, but her thoughts drifted to earlier in the day.

Top Secret. And the general forgetting his folder and me dumping my shit all over it. Christ! What is going through Leeds' mind? I was definitely not supposed to see that. Please do not think I saw that. Her thoughts ended and she dipped her forehead to her hand.

She was startled by the sound of body to chair and felt the small gust of wind that swept up as the mass of the person who sat beside her displaced the air below. She turned to see her Israeli counterpart, Rachael, looking at her in wide-eyed disbelief.

"I can't believe Professor Echud is starting on time. Ever since you Americans showed up, everyone seems to have forgotten that we generally take things 'l'at, l'at, l'at' around here." She smiled at Eva. Eva smiled back. The Israelis loved the phrase: L'at, l'at, l'at, meaning, slowly, slowly, slowly. It was the first Hebrew phrase Eva learned.

While anyone visiting Tel-Aviv University (and even a significant number of those who worked and studied there) would be blind to the scope of research then taking place, Eva was in the know. A thought flashed across her mind: *Am I blind too? What do I know?* She thought a moment. *What don't I? I know that I don't trust Leeds.* She glanced back toward the side door, the long-brewing unease Eva harbored since she first began working with DARPA began to bubble up.

Eva flipped open the green folder lying on the table in front of her and feigned interest. *Estimated immediate casualties greater than 500,000.* The line popped into her mind. *Five hundred thousand* initial *casualties? My god, what were final estimates? What are they planning?* Her stomach began to knot up. *Surely Leeds had implemented standard operating procedures to be sure that all top secret documents were gathered and destroyed. Yet that file remained behind for a period of time. How long was I gone with Cobbs?*

How could I be so stupid! She began to move her eyes around the room, imagining the multitude of places that the nanocameras could be positioned. *You're going to drive yourself crazy, Eva. What's done is done. Think. No mistakes moving forward.*

Eva suddenly realized that Rachael had been speaking to her. "What's that?"

"Are you okay, Eva?"

"I'm sorry," Eva replied, turning her eyes, but not her attention, away from her dilemma. "I guess I'm a little tired." She looked around and noticed a refreshment table at the far end of the auditorium. "I'm going to hit the restroom. Do you need anything from the refreshment table when I get back?" Rachael shook her head. Rachael began flipping through the folder materials as Professor Echud continued with his introduction. Professor Echud was a gifted scientist and speaker. He often joked that if he weren't a scientist, he would be a stand up comedian. No one who spoke to him for very long doubted he would be successful at either career. The audience was clearly enjoying his presentation and except for Eva, everyone was in rapt attention at their seats. She carefully returned the Assay papers she had been looking at to the pile on the floor and slipped out the Top Secret folder from the stack and slid it under her symposium folder. She got up and walked toward the auditorium door. As she exited she was sure to tilt the folder so that anyone looking toward her would see the back of the folder. Once out of the auditorium, she hurriedly made her way to the incinerator and deposited the top secret folder into it before heading to the restroom. She dabbed her face with a damp paper towel, but the folder's destruction did little to calm her fears. She took a deep breath and returned to the lecture.

When she sat down Professor Echud was discussing the validity of Israel as a testing ground for research involving the continuous, real-time uploading of civilian and combat soldiers' biometrics in an effort to reduce war and terrorism casualties. The idea struck her as a bit ridiculous, since Israeli lives could be saved merely by sharing PreventAll. But given America's continued refusal to do so, she supposed that Professor Echud's research was an opportunity to help reduce the needless loss of life in a county struggling to survive among many enemies.

Estimated immediate casualties greater than 500,000. She could not get the sentence out of her mind. The loss of a half-million lives would not be the result of continued terrorism or the small-scale skirmishes in which the IDF was frequently involved - it would require all-out war.

CHAPTER TEN

If I were a young man with nary a care,
I'd pack up my belongings and travel to there.
But I am an old man, well past my prime,
And here I shall remain until the end of my time.

The words popped into Clay's mind. *A song?* he thought. *Perhaps a poem.* He could not quite place it. *Sad, I remember.* His thoughts were interrupted by a dog barking in the neighbor's yard, five houses down. Grease from the pork belly cooking in a hot skillet popped and a large drop of the sizzling oil landed on the back of his wrist. He jerked his arm back from the sting, but as quickly as it came, the burning sensation left. He looked down at his hand just in time to see the red area of burned skin vanish and return to normal. The veins in his arms stood out and he gazed at the well-defined muscles. He was strong, stronger than he had ever remembered having been. And full of energy. He went to the Replicator and removed a tall glass of milk. He took a long drink. It was cold and thick, the way milk should be. He would no longer drink the skim milk of his wife's choosing, the watery fluid that only hinted at the richness of whole milk. He picked up three eggs from the counter and juggled them as he walked back to the stove where the four thick slices of pork belly were still frying in the cast iron skillet. He heard movement from upstairs. *Music. Classic Mix. Shuffle,* he thought to himself. Neil Young's voice whispered in his ear words of sin knocking, and a plea for more. He looked out the window above the sink and spotted a hummingbird at his neighbor's feeder. He watched it move swiftly from one spot to another, hovering in front of the fake flower petals and taking in some of the sweet liquid contents.

"Good morning, Dad." A half-sleeping girl walked through the doorway, across the room, sat down heavily on chair and placed her chin onto the table. Clay stared at the back of her head, wondering if she were staring out the window or sitting there with eyes closed.

"You going back to sleep right there at the table, or are you interested in a little breakfast?" he asked. *Volume down.* The music playing over the speakers softened. "I can make you some of my special French toast," he offered.

"Would you do that? That would be great," she responded, her voice hinting at a trace of enthusiasm.

"No problem at all." He still loved to cook, and he enjoyed making breakfast most of all. Generally, his daughter would make fun of him for not replicating his meals, but she knew as well as he that the Replicator was simply not capable of making his special French toast. He stepped over to the Replicator and removed some eggs and bread.

"How was the ride back home? You must have gotten in pretty late, huh?"

"Not until almost 3 a.m. We played St. Cloud last night and it's a long haul back," she said, turning her head to the side and lying it squarely on the table. Clay could see clearly that her eyes were closed.

"Yes, I know. I also know that a certain winger got a couple of assists," he paused briefly. "Way to go, honey!"

"Yeah. Thanks, but we still lost," she responded, the enthusiasm having disappeared from her voice. "Where's mom?"

"Remember? She's still out in California visiting Mark and Rich." Clay answered.

"Oh yeah," she said. "I'm surprised you didn't go. You always have so much fun with those guys."

There was a long pause. Clay broke the eggs into a bowl and started to beat in some of the milk from the glass from which he was drinking.

"When is Mom coming back?" Katie asked in a tone that lead Clay to believe she did not care much for an answer, but was simply being polite and making effort to keep the conversation going. He remembered a time not too long before when she would not have made the effort. He smiled thinking to himself, *She really is growing up.* His smile did not last long as a wave of disappointment crossed over him. *I don't even know how long she's going to be gone.* He threw out a guess. "A couple of days."

His desire got the better of him and he procured a pint of half-and-half and three new eggs, tossing the old mixture down the sink. "What have you got planned for the day?" he asked as he beat the new mixture.

"I don't know. Hannah and I are thinking of going to the lake, I guess," she groaned back.

"That sounds like fun. Another great day for it," he said. It suddenly struck Clay that recently, all the days had become quite pleasant. *What month is it?* He could not recall. *What season?*

"Can I use the car today?" Katie asked.

"Of course. You can use your Mom's. She took a cab to the airport." He poured some oil into a fresh frying pan and turned the gas below to medium-high. As the oil began to heat he dipped a slice of bread into the egg batter and let it sit. He turned his attention to the pan with the pork belly, which was starting to smoke. He turned the pork belly and set it to one side of the skillet. He then broke a couple eggs into the grease in the frying pan.

"I'm in the mood for fried eggs too, do you want any?" For as long as he could remember, he feared that his tendency to eat too much would someday catch up to him and result in weight gain, but his metabolism remained kind and he had never gained too much weight. Now he knew that the enlarging belly of middle-age would be impossible.

"No thanks." The French toast sizzled as he placed a couple slices into the hot oil. Mick Jagger's voice came over the speakers, complaining about a pain in his chest; the result of some involvement with a bow-legged woman. The French toast began to brown.

"Pork belly?"

"Sure."

He removed a few slices of French toast from the pan, patted them with a paper towel and then placed them on a plate alongside a couple slices of pork belly. He set the plate down beside Katie, her head still lying on the table, facing away from him. He moved back toward the stove and took down some powdered sugar from a shelf before turning to the refrigerator and pulling out a bottle of syrup. As he set them on the table Katie suddenly let out scream, her eyes wide open and staring at him.

"Oh, didn't I tell you?" Clay smiled. "I got inoculated. But don't tell mom. I want it to be a surprise."

CHAPTER ELEVEN

The Israeli Government continues with the universal inoculation of Israeli citizens from the Covid-19-B2 strain. As a result of their efforts, the State of Israel has experienced the lowest infection and fatality rate of any nation outside of North America.

Effra was the youngest daughter of Joseph Alfandari. As she walked up Shivat Tsiyon Street, she smiled to herself. She was nearing her twentieth birthday and was a little past the halfway mark of her two year obligation in the IDF. While she was proud of her service, she yearned to return to Kibbutz Ein Harod Meuchad and take her position in the date plantations lying below Mt. Gilboa. The fields were tended by only a few people, who rose long before most of their fellow kibbutzniks. It was not uncommon for Effra to join her friends during her mid-workday break in the common cafeteria. She would be enjoying her second meal of the day while her friends were sitting down to their first.

The work was hard, but satisfying. Winters can be tedious and boring, working on the drip irrigation systems and removing young offshoots from older trees. But the spring and summer months are spent preparing the trees for fruit, removing the spines from the lower portions of the leaves and hand-pollinating the flowers. This ultimately means working alone, climbing up and down the palms all day, like some kind of a modern-day coconut monkey. But she loved it.

For the past six months Effra had been stationed in Haifa and was responsible for providing security at one of the several natural gas pipelines transporting energy from the Leviathan gas field. She was an energetic girl, five foot six, olive-skinned and possessing long, thick, black hair. Most of the women in the IDF chose to wear their hair in loose, fetching ponytails, their fashion sense unaffected by their military conscription. Effra's insistence on wearing hers pulled tightly back and pinned up, like American women soldiers, earned her the nickname Yankee. It did not distract that her mother was an American immigrant who made aliyah in her early 20s, while working at Kibbutz Ein Harod after college. Effra possessed an easy manner that she was able to easily turn off and become a hard, determined soldier when need be. She had long ago grown accustomed to young men staring at her from a distance, but continued to be self-conscious of the initial reaction she received when she met people for the first time, face-to-face. Perhaps it was the nearness, or maybe because she could easily ignore the young man staring from across the street. But the expressions was always the same. It was her eyes that took people by surprise. They were a captivating light brown that seemed to reflect the desert sands, and provided a stark contrast to her dark hair and olive features, which usually are accompanied by similarly dark eyes. She still blushed slightly, self-conscious of her unique possession, when she caught acquaintances and friends, alike, gazing intently into her eyes.

A gas-equipment freighter, the *ben Sabea*, slowly made its way into the Port of Haifa. Effra watched its approach with apprehension. She had been assigned earlier in the day to Pier Dalet to cover for Cpl. Zev Dahon who was on leave to care for his sick mother. Effra had never worked the check point at Pier Dalet before and her commanding officer, Major Bitton, had informed her that the crew of the *ben Sabea* were scheduled to check on the operation of various equipment designed to offload natural gas. The work was routine, but critical to the operation of the Port and of the upmost importance. Recent threats from Iran had placed the nation of Israel on high alert.

"There can be no incidents whatsoever," Maj. Bitton warned. "You must remain vigilant in your duties. I cannot stress the importance of security at Pier Dalet."

The opportunity to be assigned at this juncture was a nod of approval and Effra did not plan on letting her commanding officer down.

Effra had been briefed that two times a week the *ben Sabea* would roll into at the port where its crew would work their way to the check point located on Pier Dalet in order to gain access to the main equipment room. The *ben Sabea* arrived this day as expected. After the mooring lines had been secured and slowly two men disembarked. The first was in his mid-50s, short and bald, followed by a tall, well-built man in his mid-20s. The shorter of the two walked with a slight swagger and steadily approached. The taller seemed less sure, looking back and forth, as he followed behind, perhaps conscious of something Effra had missed. The hair on the back of her neck rose and she glanced quickly around. She did not notice anything out of place and returned her gaze to the two approaching men. The shorter man was clearly of Arab descent. He had a thin build with distinctly Arab facial characteristics, his graying beard a sharp contrast to his barren head. The taller man had a ruddy complexion, but did not possess traits of any specific nationality. His apparent nervousness coupled with his continual scanning of the area raised suspicions in Effra and she swung her rifle from her shoulder to her chest, gripping both the barrel and the stock and returned her attention to the first man as they neared.

As she did so, he stopped suddenly. They stood no more than ten feet apart. He gazed intently at her, motionless. Effra moved her eyes quickly back to the taller man. "Identify yourselves," she commanded. The grip on her weapon unconsciously eased as her gaze remained fixed on the taller man. *He's beautiful,* she thought. The tall man smiled meekly and turned toward his partner. He gave the man a slight shove, but the shorter of the two remained motionless, still staring intently at Effra, whose vision remained fixed on the tall man.

"I'm Joshua," the tall man offered. "And this is Terek," he said, motioning to his partner. "We're here to perform a system check on Pump Shesh,"

Effra nodded slightly, but remained silent.

"Ma'am?" Joshua prodded.

"Of course," Effra replied, standing still.

Joshua darted his eyes back and forth between his compatriot Tarek, staring at Effra, and Effra, staring at him. "It'd be a lot easier to do the check if we didn't have to do it from out here."

Tarek caught himself staring and broke his gaze from the beauty of Effra's eyes. He let out a brief nervous laugh. "I think he's right, Ma'am. And my arms aren't even as long as his."

Effra turned her gaze away from Joshua and back to Tarek. Her face was flushed as she responded. "Forgive me. I . . I . ."

"Yeah. Me too," Tarek interrupted.

Effra let out a long sigh. It was clear to her that these men posed no threat. She stole another glance at Joshua. "Great. Day one and I'm distracted by . . ." she paused, her face starting to flush again, causing her to turn to Tarek. "I mean, I'm not making a very good impression, I apologize. I should be relieved from this post. for letting my commanding officer down. I can only imagine if you meant to cause harm . . . ." She looked briefly down, pausing once again.

"Don't worry," Tarek paused to look at the insignia on Effra's shoulder patch, "Private Alfandari. It seems many beautiful young Israelis such as yourself find my handsome and rugged looks too much. You are not the first to turn their attention to this homely man", he said, pointing his thumb toward Joshua, "in an attempt to regain their composure. But you can trust me, I am humble enough to forgive your distraction. However, I am afraid it is I who owe you an apology. Your eyes, they are as beautiful as the Sahara sands at sunset."

Effra looked up and smiled at Tarek. She glanced to Joshua.

Joshua turned his gaze quickly, and flushed himself at being caught staring into the young soldier's eyes. "And just as dangerous," Joshua added.

"Let us hope not, my friend. But I am certain that if the need arose, she could be." Tarek had noticed the subtle change in Effra, her left trigger finger now placed gently across the trigger guard, ready at an instant to move to the trigger if needed. "But enough of that," he said as he handed his credentials over to Effra. "I am Tarek Omar and this ugly duckling is Joshua Levin." He turned and smiled at Joshua. "And you, Private Alfandari, where is Zev today?"

She glanced over at the papers Tarek had handed her and set them on the table to her right. "Forgive me. I am Effra," she said, offering her right hand. Her left hand did not move from where it held its grip on the lower receiver of her weapon. Tarek took her fingers by his own and shook her hand gently. Joshua shook her hand firmly. "I'm afraid Zev's mother became ill and he had to take leave to care for her. If I'm not court-martialed, I expect to be posted here for the next week or two." She handed the documents back to Tarek.

"I see no reason why you would be otherwise." Tarek smiled.

She smiled back, and allowed the men to proceed to their work.

Thereafter, during those times Effra was not stationed at Pier Dalet, Tarek and Joshua would make it a habit of waiving toward the shore, their tiny silhouettes a sharp contrast to the blue sea behind them, unaware of whether Effra could even see them, but confident that if she did, she would know it was directed toward her. Effra would smile each time, a faint heat washing across her face, thinking of Joshua, a man she desired to know better.

Effra glanced at her watch. It was twenty-five minutes after two o'clock in the afternoon. Joshua and Tarek had come, greeted her from across the bay, and gone. The heat of the day was beginning its rapid afternoon ascent and Avi Elburg, a member of her squad, had still not returned to relieve Effra for her break. It was not unusual. Avi regularly over-extended his break, leaving Effra to cover his absence. Effra did not really mind, which may have contributed to Avi's habit of stretching his break by sometimes as much as an extra hour. At first, she had suspected that he was lazy and avoiding his duties. She quickly discovered that he was reliable and hard-working. But he also possessed a great sense of humor and an outgoing personality that made instant friends of whomever he met, and he seemed to meet new friends each day. He also was one of those people who seemed to be in the right place at the right time. Hardly a day went by that Avi did not have a story to share about some adventure had over his lunch break that prevented his timely return. He would apologize profusely and Effra had to promise her forgiveness before he would stop. She always would, not only because she enjoyed Avi and knew it was not meant in any ill manner, but because if she did not she could not leave for her own break.

Effra had never given up her habit of rising early. She, Avi and the other soldiers would often meet at a local cafe before starting their shift and listen to Avi's adventures from the previous day. While the others enjoyed their first meal of the day, Effra, as was her custom, was having her second meal of the day. Even through she would be nearly completed with her shift by three in the afternoon, it was still too early to enjoy dinner. So she used her lunch break to grab a snack and enjoy some of the many beauties of the city of Haifa. Her post at Haifa Port was only a short walk from *Hazikaron Garden*, or *Memorial Garden* as a sign near one corner notified English-speaking guests. As its name declared, it was a fitting memorial to the Israeli soldiers who died in the 1948 War of Independence. A simple park, it was oftentimes quite empty. In many cities, the park would have been enjoyed by dozens as they ate their lunch or strolled around the city seeking solace in it's natural beauty or east sight-lines. However, in Haifa, Hazikaron lies, as does much of the city, in the shadow of The Gardens, better known as the Bahá'í Gardens, which rise to the top of Mount Carmel and draw over one million visitors a year. Hazikaron itself contains an open, grassy area, surrounded by a stone walkway. A simple fountain bubbles on one end, backed by trees. The opposite end contains a raised stone patio with benches looking to the Mediterranean Sea. Effra saw enough of the Mediterranean as she worked and would venture towards the fountain and sit among the trees, her back offered to the sea. But it wasn't just the familiarity with the Mediterranean that prompted Effra to ignore it. The truth lay in the fact that right across the street, the children of the Benjamin M. Safrad Day School would be enjoying their afternoon recess. Effra loved to watch as the children chased each other around the play area, laughing with the joy that only small children who are ignorant of the realities of the world can truly embody. And, of course, she always enjoyed her visits with Noa.

Effra had met Noa about four months previous. The day was bright and the sun was sharing its warmth with anyone willing to accept it. Effra sat on the bench, tearing a small piece of pita from the whole, watching some young girls at Safrad Day School skipping rope and singing their simple song to keep time. From the corner of her eye, she spotted a red ball, coming from the direction of the schoolyard, rolling toward the street. As she stood up to retrieve the ball she spotted a young girl, no more than five or six, following behind, ignoring the traffic as she chased the ball towards the street, a pink rabbit held tightly in one arm. Effra turned her head from the ball as it continued towards the opposite curb, its progress stopped by the tire of a parked car. She shouted out to the young girl.

"Stop!"

The little girl raised her head and froze in place, between two parked cars only inches from the street. She stared intently at Effra. The rabbit dropped from her side.

"Don't you know you could have been hurt?!" Effra continued to shout. She trembled slightly as the adrenaline continued to race through her body. She took a deep breath. *Thank you, God,* she prayed, trying to calm her nerves. The little girl, still standing motionless, realized her error and began to sob. It melted Effra's heart. She carefully looked both ways before stepping into the street. She made a slight detour to retrieve the ball and spoke, this time in a gentle voice, "Don't cry. It's okay." She held the ball out to the little girl. "I've got your ball." The girl lifted her head, still sobbing. Effra tossed her the ball while standing in the middle of the street. The sound of a car horn caused both Effra and the little girl to jump. Effra turned to see a car slowing down to avoid hitting her as she stood in its path. As she moved closer to the curb, Effra raised her hand in apology to the driver, who shouted a few colorful expressions, and then Effra turned to the young girl with a grimace on her face and, hamming it up a little, jumped towards the curb. The little girl laughed a bit as she wiped the tears from her face. That's when the little girl adopted Effra.

A teacher from the school who had been in charge of supervision came around the corner of the schoolyard fence.

"Noa, Noa! Oh, dear, are you okay?"

"Yes, Ms. Sophie," Noa whispered.

She placed her hands around the girl and tried to lift her into her arms. The young girl squirmed and Effra took a step closer and picked up the rabbit lying near the curb. She handed it to Noa, who grabbed it and then allowed herself to be picked up.

"What were you thinking, Noa? I was so afraid you were going to step out into the street," she chided before turning her attention to Effra. It was clear from the expression on her face that Sophie did not know fully what had transpired and was in the awkward position of not knowing whether to scold Effra for some unseen transgression or to thank her for protecting the young girl. In either case Sophie was aware that she had been remiss in her duties and, apparently thinking it best to not draw attention to her own failings, simply turned around.

"Oh, Noa. It's okay dear. You're okay," Effra heard the teacher say as she rocked the still sniffling young girl gently in her arms. This time it was Effra who remained motionless on the curb. The woman did not look back and Effra watched them disappear around the fence and into the school. A bell rang, and Effra, still standing there, watched as all the children hurried back inside. She finally turned back to her snack. A squirrel was sitting on the edge of the bench, finishing the pita. It spotted Effra and dashed off, leaving Effra with nothing but a few crumbs, and the lingering picture of Noa in her thoughts.

The following day, as Effra approached the bench she usually occupied to enjoy her snack, she spotted the little girl waiting by the schoolyard fence, waiving, her pink rabbit safely by her side. Effra smiled as she walked past the bench and across the street. She approached the fence and bent down so she would be eye-level with the girl.

Effra held out her hand, guiding it between the iron slats of the fence. "Hello. My name is Effra."

The little girl immediately reached for Effra's outstretched hand and grabbed it. "Hi. I'm Noa. You're the lady who got my ball yesterday. I was hoping you might come back." She smiled and looked down, a shy grin flashing across her face. She pulled her hand from Effrat's and made a small curtsey. Effra stood up and pulling on the sides of her military trousers, returned the curtsy.

"Pleased to meet you, Noa." Effra smiled a wide grin.

Noa began to immediately start talking about her day at school and her friends in class. Effra could tell that Noa had no interest in even taking a breath to pause her story and so she quickly stood up to catch sight of a teacher or other adult supervisor on the playground. She motioned an elderly woman over and bent down to Noa to explain that she would be considered a stranger to the teachers and that she wanted make sure that it was okay for the two of them to sit and talk. The elderly woman approached and Effra explained how she had met Noa a day earlier and asked if it would be a problem if she sat outside the playground fence and ate while talking, or more likely, listening, to Noa. The teacher asked for identification and asked a number of questions of Effra before determining that she presented no harm to the child and consented. Thereafter, Noa would wait by the schoolyard fence each day for Effra to come with her lunch and they would sit and talk while Effra ate. This pattern continued every day, except the day Avi failed to return from his break. Major Bittin later described how he had been injured and Avi killed attempting to prevent a terrorist attack against a civilian target in Haifa.

* * *

When the recess break began and the doors were opened, Noa ran out to the fence. Sophie smiled to see the little girl dash out the door. It had taken some time for her to acclimate to her new surroundings and her relationship with Effra was a catalyst in Noa's adaptation to life in foster care. As her relationship with Effra grew, and she learned to trust Effra, she seemed to allow that trust to carry over to her new surroundings and her classmates. Sophie looked out toward the park, where Effra would be eating her snack, waiting for the children's recess to begin. A frown grew on her face. Effra was not there. She glanced back at Noa, who realizing that her friend was not present, stopped suddenly in her tracks. Her shoulders dropped. Sophie walked over to where Noa had stopped.

"It's okay, honey. I'm sure she's just a little late today. She is a soldier and sometimes she has other very important things to do. Why don't you sit near me and we can wait for her to get here together."

Noa nodded and sat down near Sophie's feet, never taking her eyes off the empty park bench.

* * *

For the third time in as many minutes Effra asked her iMeme for the time. Avi was often late returning from his break, but never this late. Recess would be ending at the Safrad Day School. It would be the first time Effra did not visit Noa since they met several months earlier. Her heart dropped as she thought what Noa may be thinking, but her thoughts were soon darkened by worse thoughts. *Where is Avi? Is he okay?*

Break had started out uneventfully for Avi. He was hungry and decided to hit a local street falafel stand. As he was exiting the base, Major Bittin stopped him. The Major was accompanied by two U.S. Soldiers. Avi saw by their insignias that the short one was a colonel and the larger man a sergeant.

"Private Elburg," the Major began, his voice unsteady and his gaze staring beyond Avi. "These Americans have asked if you could join them momentarily to assist them in some . . . er . . . confidential matters." Major Bittin glanced nervously back at Colonel Leeds. Avi noticed that Bittin had a bruise growing around his eye.

"What is this about?" Avi demanded, taken aback by the Major's appearance and sensing tension. Recognizing that he clearly overstepped his bounds, he quickly spoke again. "Sorry, sir! I mean, is this a voluntary request, Sir?" Avi broke into a sweat, unsure whether the major wasn't being forced into his request. He eyed the American soldiers, uncertain of whether they were, in fact, who they claimed. His body tensed, ready to take action if necessary. "I was just preparing to take my leave, Sir, and if it is not necessary." He struggled to assess the situation and sought of some way to give Major Bittin an opportunity to offer a hint as to what was going on.

Sgt. Mueller noticed Avi's reaction to the situation and grabbed his arm. "It is not a request, Private. I'm certain it is an order from your commanding officer." Mueller directed a false smile toward Major Bittin. Avi tried to pull his arm away, and was surprised by the strength of the Sergeant. At that moment, Major Bittin shouted.

"Let him go soldier! Avi, don't-" Major Bittin was interrupted by Sgt Mueller's fist. Mueller landed a solid blow to the major's already bruised eye without loosening his grip on Avi's arm. Major Bittin fell to the ground. Avi stopped struggling, shocked by the events.

"I'm sorry Major if I did not make myself clear," Leeds said, looking down on the crumpled body of Major Bittin, who was attempting to raise himself, rubbing his now bleeding eye. "We will be taking this soldier." Leeds turned his gaze toward Avi.

"Private Elburg. I hope my Sergeant has made it clear by now that this is not a voluntary request." Leeds gave a nod towards Mueller. Avi began to struggle again and Mueller removed his sidearm from its holster and brought it down on Avi's head.

"You can't—" Avi started before everything went black.

* * *

"Please do not struggle or attempt anything stupid. We are testing a new technology and you, Private Elburg, are the man who may help prove one of the greatest technologies ever created for war.

"Sadly, you will die, but only momentarily."

Avi's eyes widened in fear.

"If you are cooperative, that death will be painless. Should you choose to struggle . . . well . . . ."

Leeds looked at Mueller. Mueller grinned. Avi lay stone still, strapped to a gurney. Fear had stricken him. He was unable to move even had he so chosen.

"Good," Leeds continued. Mueller frowned as Leeds removed a syringe from the instrument stand next to the gurney and injected Avi. Once again, for Avi, everything went momentarily black.

Avi blinked. *What was that?* He looked around, his voice coming back to him. "I . . . I . . . I don't understand. Is this some kind of joke?" He began to struggle against his retrains.

Leeds was smiling. "What is the last thing you recall, Private Elburg?"

"You injected me with that needle and it seemed like things went black, but then it didn't. What's going on?" Avi demanded, ignoring the possibility that Leeds was being truthful when he offered a painful death.

"You have passed the test. Congratulations. But I do have one more question for you. If we were to run into each other some day in the future, would you recognize me?"

"What? Of course. Why do you ask?"

Leeds smiled and turned to Niklas. "Kill him."

CHAPTER TWELVE

Iranian military generals have renewed calls for the total destruction of Israel and the return of Palestine to the Palestinian people. They have also indicated that they are normalizing relations with North Korea and have threatened to send troops to support what it has characterized as "hostile intensions" from the South Koreans to their friend and ally.

"Mozel Tov, Eva. You've done it." Rafi smiled as he tipped his plastic champagne glass and took a large sip of the bubbly contents.

"Congratulations, Rafi, *we've* done it," Eva corrected him, feigning a sip. Although she was excited for her friends, she had too many other matters occupying her mind to dare entertain the notion of introducing the overly-sweet kosher offering to her stomach and bloodstream.

"I did not forget about you, Eva. Here, put that down," Rachael said as she gently removed the plastic champagne flute from Eva's hand and placed it on her desk. She then handed Eva a coffee mug filled with liquid. Although the dark color of the mug hid the true color of its contents, it was not the color of the dark sludge that the Israeli's affectionately called Turkish Coffee, or worse, the oft-shunned Sanka: even captured in the darkness of the bowl that was the mug, Eva was certain of that.

Eva brought the mug to her lips and took a cautionary sip. She struggled to avoid an obvious pucker as she felt the salivary glands in her lower jaw tighten.

"It is good?" Rachel asked as she revealed a pitcher of yellow-greenish liquid she had been hiding behind her back." A little something special for you."

"Mmmm, Margaritas. How thoughtful," Eva forced a smile. "You must have one."

"Oh no, Eva. I will stick to the champagne. I do not like much for tequila," Rachel said as moved her body away from Eva slightly, and took a large drink of her champagne. She poured herself another glass as Eva wondered where the bottle came from.

It's no wonder, Eva thought, *if this was your first taste.* Eva swallowed hard in an effort to remove the lingering tartness from her palette. "Quite good, Rachael," Eva lied, "but it's usually served on ice.

"No problem," Rafi said. He had been standing behind Rachel, hoping for a taste of the margarita. "Professor Abramson has some dry ice in his lab. I'll run and get some," he said as he turned to leave.

Eva had learned almost immediately after her arrival that the Israelis did not seem overly fond of ice cubes. The urge to clear her mind suddenly overtook her better judgment and notwithstanding the taste, she took a long draught from the mug, nearly emptying it. *God,* Eva thought, *I'm back in high school drinking booze stolen from some parents' liquor cabinet. 'Oh, don't worry what it tastes like, just drink it!'*

Eva looked thoughtfully at Rachel.. "Thanks, Rachel," she said as she once again struggled to stop the involuntary urge to pucker her lips. The cause of her current torture was, of course, her own fault.

You've never had a margarita? Eva recalled asking Rachel some time back. *It's tequila and lime juice. Delicious! And it can pack a punch. I love them.*

At the time, Eva felt no particular need to mention Cointreau, Triple sec or at least a little simple syrup or sugar in the list of ingredients. *Thank God I don't like a salted rim,* she thought, as the saliva glands in her lower jaw began to feel like were convulsing in a maddening attempt to squeeze out all their contents.

"Can you believe how good the presentation went?" Rachael boasted. "And Colonel Leeds, he seemed very happy." Rachael gave Eva a gentle elbow to the ribs and Eva noticed that Rachael's eyes were beginning to glass over and get that faraway look that meant the alcohol was beginning to take its desired effects. "I think he likes you, Eva."

Eva took a quick sip of the libation in her mug and was glad for the urge to pucker. It hid her expression of revulsion at the thought of her and Col. Leeds being looked upon by the others as a potential couple.

"While I respect the Colonel very highly, there is absolutely nothing going on between the two of us. The idea of that would be-" Eva paused, trying to capture the right phrase to convey both her dutiful respect of Leeds as her boss (and a man she did not fully trust), and her desire to make it clear that such a thought should never cross Rachael's mind again. She took another large drink from the mug, finding the flavor less unpalatable than before. *Hmmm. How much tequila is in this?* As she moved the mug away from her lips she saw the Rachael had already finished yet another glass of champagne and was very clearly advancing from happy buzz to drunk.

Rafi returned wearing two large cryogenic gloves and dropped a small piece of dry ice from his left hand into the pitcher of margarita and an even smaller piece from his right hand into Eva's mug. "Here you go."

Rachael smiled at Rafi and swayed slightly. Eva noticed that Rachael had started the slight orbital stance of one who is on the verge of having had too much to drink. She felt no need to complete her response to Rachael's comment or to risk saying something that could come back to Leeds.

Rafi looked at Eva and winked as he tilted his head towards Rachael. "Not much longer now."

Rachael was smiling and humming to herself as she staggered over to where Lars was sitting alone. She turned towards Eva and Rafi and gave them a knowing smile.

Rachael generally enjoyed harassing Lars. "This should get interesting, given her current state of mind," Eva said.

"I am ready to watch the fireworks," Rafi said. "The dry ice seems to have boiled away," he said as he helped himself to a taste of Eva's drink. He let out a cough. "Oy vey! You like these margaritas drinks?"

"This is not a margarita," Eva responded. "But have a little more. It gets better," she assured him.

"Lars!" Rachael shouted as she placed her arm around his shoulder. "Lars, my friend." Rachael seemed to stumble a little and let out a small giggle. "You were fabulous today. It seems your mathematics actually do work. I am not afraid to say that I did not have confidence in you." Rachael looked over towards Eva and Rafi and winked.

"Please Rachael," Lars said as he removed her arm from around his neck. "I do not wish to be subject to your insults. This is meant to be a celebration of our accomplishments. This lab showed today that we can upload real-time biological data and download it into human biometric simulations to develop quick and effective diagnosis to a variety of physical ailments, and institute treatments to help save lives. While you may feel that my contribution was less significant than yours, I can assure you that Col. Leeds was most impressed. Did you see his reaction when information relating to the induced atrial defibulation was transmitted to the projection? Providing him with information not only on the medical crisis, but also the identity of the particular laboratory subject. While the sample study size consisted of only ten subjects, I promise you, the sophisticated formulas necessary, and which *I* developed, would work across large numbers of subjects. Larger, I may suggest, than the number of troops currently comprising the IDF, and more than capable of handling ten times that amount!"

"She's just giving you a hard time, Lars," Yitzak said, handing Lars another glass of champagne. "Really, we know your contribution was important to our project. I mean, not everyone can teach a computer to count", he said sarcastically. He smiled and went on. "But seriously, I am only joking with you. Your work was fantastic," Yitzak said with real sincerity in his voice. "You said it yourself. I have not seen Col. Leeds so seemingly happy before."

* * *

The conference had started innocently enough. The multiple lab groups were presenting a real-time demonstration of the technology required to upload biodata. In addition to the various academics invited to the presentation, there was a group of top American and Israeli military brass. Eva knew that Col. Leeds would be attending with several of the Niklases. However, the presence of so many other military personnel set her at unease. The top secret folder leapt to the front of her mind and she felt herself begin to perspire. She had little doubt these men were present at that meeting. Unlike most in the room, she knew that the ability to access biometrics was not the sole purpose of the work being performed at the Tel-Aviv University labs. But she was not foolish enough to think that she knew the real purpose for Col. Leeds having brought her out to Israel to research what amounted to high-tech communication matters.

The soldiers dominated the center parts of the first three rows of seats. Seats in those rows not occupied by soldiers remained vacant. Eva was scheduled to have a seat at the presentation table on the stage. She was not a lead investigator (for appearances' sake, she was in Rachael's group) but understood that her status as the only American scientist working on the project afforded her the luxury of special treatment. The presentation table was set with name cards for the various scientific staff tasked with detailing the progress of the work. *Thank God,* Eva sighed as she saw her name card near the far right end of the table. On her immediate left, in conversation with the woman seated to his left, was Dr. Avram. An elderly scholar who provided valuable metrics for developing the source code many of the lab groups were working on, Dr. Avram had a habit of nodding off during large events due, in part, to his inability to discern voices among the many audible background noises of a large group, thus preventing him from actually following the presenter. In an attempt to hear better, he would close his eyes in order to concentrate on the speaker. More often than not, however, it resulted in him falling asleep. The seat to Eva's right had no name card. Eva sat down on the chair and leaned back. The chair had a certain amount of give in the seatback and it was much more comfortable than she had expected.

Dr. Wallach who was sitting at the center of the table stood up. "Shalom. And welcome to our IDF and American military friends," he started. The room quieted down as those present took their seats or turned their bodies to face the stage. The warmth and gentle buzz made her feel sleepy and she considered the consequences of joining Dr. Avram in his inevitable nap, but thought better of it. She let her mind wander and it moved to the top secret memo as she carefully studied the faces of the various American generals present. Like all Americans, their appearance likely belied their age. Eva had always found that the eyes were able to give a hint of their owners true age. Of course, like the rest of one's rejuvenated PreVentall being, the eyeballs, themselves, were no less young. But the expression of knowledge, the experience gathered through one's years added a certain light to one's eyes, a certain unquantifiable change of which Eva could usually espy. It surprised her, then, that these soldiers seemed free of the earlier burdens of their lives. Their eyes shone bright, not with the sparkle of knowledge, but with the inexperience of youth. She wondered if, perhaps, these soldiers were truly young men, promoted in a system that most back in America found to be irrelevant and, frankly, archaic. *Who, after all,* Eva thought, *would ever consider war with America? War against a country whose soldiers could not die? Whose property could not be damaged? Whose ability to rain death and destruction could not be imagined?*

"Oh," Eva blurted out in a muted surprise as she felt an unexpected touch at her right shoulder. She turned quickly and Colonel Leeds smiled at her as he removed his hand and took a seat beside Eva.

He leaned over, still looking out at the audience and whispered. "I am looking forward to the latest results of your work, Professor Diaz."

She forced a faint smile as she turned toward Dr. Wallach. *Why isn't he sitting with the other officers? Odd,* she thought. But she knew it was not odd at all, the moment the thought flashed through her brain. The conference was a demonstration of the advanced biometrics system meant to provide soldiers the best chance for survival in a military operation. The American and Israeli military officers were present for very different reasons. The IDF was seeking methods to ensure the safety and survivability of its troops. The American officers clearly had no real need for the technology given that all American troops were all inoculated. *Perhaps,* thought Eva, *it was concern for fellow soldiers. Perhaps it was simply a smug curiosity to observe how the other half will conduct itself in war and tragedy, much like small children will sometimes wreak havoc on an ant nest to watch how they react to disaster.* She looked once again into the eyes of the U.S. officers seated in the audience. *No,* she concluded. The dullness emanating from their gazes betrayed their absolute ignorance. *It's a farce. Those men are simpletons. The kind of men who can sit and listen for hours to the teachings of others, but who were incapable of forming a single question in response, of possessing any original idea to confirm or challenge the topic presented. Patsies, really, who will nod and smile and feign interest throughout the presentation, yet would walk out knowing no more than when they walked in, and, worse,* thought Eva, *indifferent to their ignorance. Never considering for a moment anything other than where they would have a drink after the conference and, perhaps, with whom they could manage to woo into bed.*

It would not occur to them that America's interest in assisting the IDF in biometric development was simply a cover for other technology. Eva looked back over to Rachael and her team. *They are Israelis and - like all Israelis - having had served in the IDF they must certainly be aware of the benefits of the technology we were working on. Maybe that's why they never really questioned me. They are far too savvy to risk jeopardizing the benefits of biometric assessments by asking me too many questions. And for that, they will let the underlying work go unquestioned.*

Just like me. Eva closed her eyes. *I am no different. Am I really trapped in this job? What would happen if I refused Leeds' requests?* She shuddered for a moment. She knew that was not a possibility. *Who knows what he would do to my family, to me.*

She thought back to the last time she had seen Clay. *Why did he come to mind?*

The lights in the auditorium dimmed as Professor Wallach began his PowerPoint presentation. Eva was not listening to what he had to say and welcomed the relative privacy the darkened room offered. Her thoughts continued to roam. *Had I said too much to Clay? Did I put him in jeopardy?* She was unsure of what Leeds may have known, but there was no doubt in her mind that Col. Leeds was a dangerous man. And obedience to his desires was, seemingly, the only choice.

She glanced over at Colonel Leeds. His face showed no sign of emotion. Neither the disappointment he expressed when they last tested the progress of Eva's work, nor the faint amusement at the first. Eva revulsed at the memory. The multi-way communication between and among the various Niklases was easy enough to accomplish. The continuous uploading of the various new, modified, changed or stimulated synapsis in each Niklases brain used the same technology developed in the biometric work being performed at Tel Aviv University. The trick came in delivering those changes to each and every other Niklas.

Niklas Beta 1 was a subtle failure. The theory behind Beta 1 was to transfer stimuli inputs from the other Niklases directly into Beta 1's frontal lobe. While Beta 1 did acquire the thoughts and memories of the various other Niklases, it amounted to little more than vague memories of the experiences of the other Niklases. There was no immediate knowledge, merely an intermittent understanding that generally only could be elicited by careful questioning or other methods to trigger the memory response. Naturally, this was not acceptable to Leeds. Beta 1 was little removed from almost any other Niklas out there.

Eva settled deeper into her seat and scanned the audience. All eyes were trained on Dr. Wallach. She closed her eyes. Leeds' last words echoed in her mind. *How is Leeds going to test Beta 3.* Her eyes shot open. *My god! Is he going to be testing it here!* She turned suddenly and glanced over her shoulder back to the Niklases. *Are one of these two Beta 3?* she wondered in horror. Niklas Beta 2 was a complete failure.

* * *

"Are you ready?" Leeds asked the Niklas.

Eva's stomach turned. The Beta 2 test was devised almost entirely by Dr. Ho. Eva had voiced her objections to Colonel Leeds, but he was insistent on moving forward. Beta 2 would receive the real-time biometric data of all the Niklases directly into his brain, completely unfiltered. Dr. Ho argued that it was necessary to determine whether real-time temporal input was even possible or whether the human brain would retain the synaptic changes without actually integrating the changes into the normal brain functioning.

"Sir, yes sir!" The Niklas snapped.

Leeds turned to Dr. Ho, who had arrived in Tel Aviv earlier in the day to observe the test for himself. Ho slowly nodded. He did not even glance over to the Niklas as he entered the code into the computer system. Eva turned away from the Niklas who would later be known as Beta 2 and her gaze landed on the two Niklases standing near the doorway. She was struck by the notion that a man with seemingly no original thoughts was about to receive the unoriginal thoughts of his hundreds, perhaps thousands, of other unoriginal selves. His gaze was steady and his face expressionless as Dr. Ho completed the sequence and straightened his body, finally looking to Beta 2, as his brain began receiving the transmissions from the other Niklases.

Beta 2's eyes instantly grew larger as he stiffened. His whole body seemed to begin to vibrate, almost imperceptibly at first, as he crashed to the floor in a grand mal seizure. Eyes still wide, Beta 2's head began moving rapidly, side-to-side, the reflection from the blinking lab computer lights causing the floor around him to glow in a myriad of colors. Eva was struck by the image of Dr. David Bowman as he met the Monolith and raced toward his rebirth as the Starchild in Stanley Kubrick's adaptation of Arthur C. Clark's *2001: A Space Odyssey* onto the big screen. *What is the human mind capable of?* she wondered. *Is this how the body reacts when looking through a thousand sets of eyes, listening with a thousand sets of ears and sensing the tactile stimuli from a thousand different inputs?* Beta 2's eyes rolled into the back of his head, his mind unable to decipher in which of the many realties it was experiencing his physical being belonged. His body began to convulse more violently, struggling to move in a universe in which he felt the multitudes of floors and stairs and seats and beds and environments that did not match his physical surroundings. He opened his mouth slightly and from the space somewhere beyond his lips a faint, inhuman, sound arose. Low at first, it grew louder and higher in pitch as his body writhed, his arms and legs flailing about like a man falling through the air. A small stream of blood began to flow out of his nose. Eva moved towards the struggling man, but Leeds held her back. Blood now began to flow from both his ears, the sound emanating from Beta 2 so high-pitched it began to hurt Eva's ears. She turned her gaze away from the horrifying scene when, suddenly, there was silence. Beta 2 lay motionless, a pool of blood forming beneath his head. A small grin crossed Leeds' face and as he released Eva. She immediately fell beside the soldier, the blood seeping into the knees of her pants.

"He's dead," she said as she turned back towards Leeds, her mind racing to comprehend how the PreVentall failed.

"Clean this mess up," he said. The grin had not faded from his countenance. He grabbed Dr. Ho by the arm and the two proceeded towards the exit, whispering in private discussion. The two Niklases stood motionless as Leeds and Ho left the lab. As if Leeds' departure were a signal, without a word one of the Niklases picked up the body of his dead duplicate. Eva remained motionless, numb to the efforts of the other Niklas as he knelt down beside her and began the process of cleaning up the sticky pool of bodily fluids.

* * *

Eva sat more upright and concentrated on not looking at either Leeds or the Niklases. Eva's reconfigured test, Beta 3, was designed to avoid the deliberate placement of stimuli inputs into the frontal lobe. Rather, it would transmit and cause each new or changed synapsis connection in the cerebrum of every other Niklas brain to be duplicated in Beta 3's brain, specifically excluding the brain stem and, thus, avoiding the direct stimuli inputs experienced by the individual Niklases, and what caused the sensory overload in Beta 2 that was so intense it shorted out the PreVentall nanobots. The concern, however, was what would happen when the identical synapsis was changed simultaneously in two separate Niklases. The actual outcome was uncertain. A growing fear that Leeds was ready to run the Beta 3 test was growing inside her. Her mouth suddenly went dry. She looked around, wishing she had brought something to drink. There was a pitcher and several glassed near the far right corner of the table. She could not force herself to turn towards Leeds to whisper the favor. She closed her eyes and took a deep breath.

When Eva re-opened her eyes she found herself staring at Rachael, Yitzak, Rafi and Lars, sitting nearest the center of the auditorium without violating the space reserved for military personnel. Her discomfort in sitting next to Col. Leeds, the two Niklases standing behind him, increased as she feared the worst.

Another gentle touch of Col. Leeds' hand on her shoulder interrupted her thoughts. She jumped in her seat and for the second time she let out a muffled *Oh*!

"I apologize, Doctor," Leeds whispered, leaning over toward her ear. "But I am certain you must be curious to see the latest, and hopefully final, fruits of your efforts?"

Eva did not know what to say. She gave him a look of mild confusion to which he simply smiled as he leaned back into his chair and motioned a Niklas to his side. He whispered into his ear and almost immediately a third Niklas appeared to Eva's right. He quickly and quietly stepped behind Col. Leeds and leaned over and whispered quietly into Leeds' ear. The expression on Leeds' face went serious, and as he moved his chair back from the table in order to stand up, he looked once again over to Eva. She noticed a twinkle in his eye.

"Let's see if our little 'test' works," he whispered quietly as he stood up and exited from where the third Niklas has recently come. All three Niklases followed. Eva was watching the audience who was wrapped up in Dr. Wallach's presentation. No one so much as looked over at Col. Leeds as he departed. Not even, Eva noted, the U.S. officers seated in the center of the room.

CHAPTER THIRTEEN

It is Tuesday.

The house was quiet when Lillian returned from her trip to the west coast. "Hello! Anyone home?" she called out.

"I'm up here," Clay answered.

She went upstairs and found Clay lying in bed, covers pulled up around him. "Are you okay, honey?" She was getting onto the bed to put her arm around him. He rolled over to his back, smiling, and pulling the covers off from over him. His naked, tanned body stood out against the white sheets. His muscles were firm, a hinting of a six-pack expressed as he raised his body slightly. His pectoral muscles were pronounced next to his broad shoulders and round deltoids.

"Oh my fucking god! You did it! You really did it!" She smiled as she looked up and down his body over and over. "Jesus, you look great!" He smiled back at her. She reached out and ran her hands down from his shoulders, across his chest, to his legs. She moved her hands up slowly and stopped. She leaned over and began to run her tongue slowly over his body, red lips parted slightly, while removing all her clothes. She paused briefly to remove her shirt. She smiled at him as she turned around and climbed onto the bed, her legs straddling his head. She continued to tease, darting her tongue all over his firm body while lowering her hips closer to Clay's face. He pulled her close and kissed her before he rolled her over and turned around to stare into her eyes. Today there would be no toys or virtual experiences. They made love, the two of them, living, breathing and real. When it was over, Lillian lay on her back, staring at the ceiling, smiling. "God, I missed that. I missed you. The old you – no, no, not the old you, the you, *you* used to be a long time ago, the young you. The you, you are now." Clay laughed. "Shit, I don't know what I'm saying. I love you, Clay. And I'm glad you're young again. Thanks. I know you didn't want to do it, but I hope this helps make up for it." She rolled over and kissed him passionately on the lips. He pulled her on top of him and wrapped his arms tightly around her, returning her kiss. They continued to kiss as he rolled them both over so he was lying on top of her. As he nibbled her earlobe and kissed down her neck, he felt the urge to make love to her again. And he did, thinking that, perhaps, youth really does have its advantages.

CHAPTER FOURTEEN

> *The time is fast approaching. Have not you been devout in your service to me? Have not the People of the Republic? The enemy of your people no longer speak of Wang Geon; your Southern cousins have forgotten their duty. Their service is no longer worthy of the deities. You shall prepare your armies. The Gunungshin have decided that victory shall be yours!*

Kim Yo-Jong awoke in a cold sweat. She had grown accustomed to the advice of The Great Leader from her dreams. But the Gunungshin? They were not merely eternal, they were deities.

"I am listening, Wang Geon" she spoke aloud. "You shall be praised throughout the Kingdom as the greatest of the Gunungshin."

There were no more words. She stared at the ceiling, smiling to herself as she imagined her soon to be realized good fortune. She turned her head and saw a young woman lying there, breathing rhythmically. *Did I not lie down earlier next to my husband, Choe Jong?* She pulled the covers over herself and rolled closer to the unknown woman. The young woman awoke with a start. Faces close together, the young woman smiled at Kim Yo-Jong, but her eyes showed no evidence of joy. *I do not care.* The smile faded. "You do not matter, woman," she spat. She wondered where Choe Jong had gone to, what he may be doing this night, but her thoughts quickly returned to Wang Geon. *You do not matter, either, Choe.* "Only Wang Geon matters now," she whispered. She drew closer and bore her tongue deep into the young woman's mouth, as she forced her hand between the young woman's thighs. The young woman tensed briefly before she submitted and spread her legs, but Kim Yo-Jong had felt the resistance. She removed her hand and slapped the young woman on the cheek. She sat up and straddled the young woman's face. She adjusted her knees and removed the pillow lying beneath the young woman's head and threw it to her right. It struck the American soldier who was standing beside the door in silent attention, facing forward, his eyes likely in rapt attention to nothing in particular at all. Ever since the Americans arrived Kim Yo-Jong's fortunes had changed. She threw a hard glare at the soldier. *Soon, you too, Sgt. Mueller, will answer to me. You and Col. Leeds. I have the deities on my side.* A smile crossed her face as she continued to gyrate on the young woman below. A thrill that began as a physical response radiated through her body, rising as it mingled with her imaginings of power; her hips rocking faster and faster as the thought of her invincibility grew in her mind. For the first time, Kim Yo-Jong felt something beyond mere physical pleasure from sex and believed that she had, at last, become one with the deities. She looked down, pitifully, at the young woman below, who made no sound, expressed no emotion. As a woman who was born into absolute power, Kim Yo-Jong did not know any better. She considered the disconnect between the fury of intercourse and the lack of emotion as normal. Like an animal, she had no notion of sex as anything other than a physical act. She let out a scream as her body trembled and rolled off the woman.

"Thank you, oh Most Supreme Eternal Leader," the young woman choked, gasping for breath. She quickly wiped off her mouth and added, "You are most excellent." The young woman feigned a smile and rolled to her side.

Kim Yo-Jong's eyes opened wide, her mental state acutely aware. *Who is this woman?* She turned toward the woman, whose back was indistinguishable from any of the possible men or women who could be lying there. The idea flashed through her mind again and again: *The Eternal Leader.* She smiled to herself. *I will surpass you Great Leader, and you too, Dear Leader. I am the Eternal Leader!* She leapt from the bed and turned toward the woman lying there. "Be gone woman before I return," and exited the room, spitting on the soldier still standing at stiff attention at the door.

"Bitch," he mumbled as he wiped the spit off of his shirt.

"A small price to pay, Sergeant," Col. Leeds' voice responded from the throat of the young woman, who rose from the bed and wrapped the sheets around her naked body.

"Sir," he said snapping to attention. "I did not realize it was you."

"Yes," Leeds smiled. "It's new technology. The nanobots can cause light to reflect off your skin in such a manner as to present yourself as whomever," he paused, "or whatever, you like. It definitely has its distinct advantages. You were not aware of this? he asked.

"No Sir, Colonel."

"That is good. Dr. Ho has assured me that this ability to manipulate the 'bots in our cells into modifying their reflection is not available to anyone but me. I don't believe him, but I expect that the next PreVentall upgrade will provide sufficient built-in security to make it nearly impossible for anyone but the most sophisticated to succeed in such a hack, if the thought even crossed their mind. And since anyone capable of such a hack all work for me, I am willing to ignore his untruths. For now.

"But the pressing issue is that of our lovely Ms. Yo-Jung's mental image of Wang Geon. It is easy enough to plant dreams of Wang Geon in her head, but in order for my plan to work, I must come to her as Wang Geon. In the flesh.

"Have her dreams extracted and images sent to Dr. Ho. I want to be certain to get this right. I'd hate to have to kill Ms. Yo-Jung. We've come so far and our little North Korean friend is such an amusing character." Leeds let the covers slide off his body as he followed the path Kim Yo-Jung had just walked.

Sgt. Mueller stared at the naked backside moving down the hall and laughed to himself. *The bots may make him look like a woman, but he still walked like a man.*

CHAPTER FIFTEEN

The Israeli Ambassador to the U.N. called for an emergency meeting of the Security Council to address Iran's continued threats of annihilation against Israel. It is expected that little will be accomplished at U.N. Headquarters in Brussels, given the Security Council's failure to take any action after evidence was presented over three years ago that Iran had failed to comply with both the Joint Comprehensive Plan of Action and the Trump-Iran Plan and had, for a number of years, been running more than 20,000 centrifuges at Natanz. Along with growing circumstantial evidence of on-going nuclear development, including the apparent close cooperation between Iran and North Korea, the Israeli Ambassador had hoped to slow the ever-growing body of anti-Israel actions by the world body. However, American indifference to U.N. proceedings, except to advance American agenda, has resulted in an increasingly belligerent atmosphere for the Jewish Nation.

Eva kicked her legs up onto the couch and placed her head on the oversized arm. The couch was old and Eva wondered if it had ever had sufficient padding to lie without the need of a pillow to soften the wooden frame beneath. She was torn between the effort to get up and retrieve a pillow from the chair across the room, or the convenience of remaining, head aching from both the discomfort of the hard surface beneath it and the over-consumption of cheap Israeli tequila. She closed her eyes, but quickly reopened them when the mild sense of spinning fell upon her. *Hydration,* she thought. Her body begged her to remain, motionless and uncomfortable, as her brain tried to coax her into sitting up. Her eyes closed again and she concentrated on the back of her eyelids, on keeping the world still. As the she began to feel the earth spin again she tried to take her mind off of the uneasy tilt-a-world feeling. *What the hell did it mean, more than 1 million casualties? Operation Phoenix . . . .*

Eva was still staring at the back of her eyelids. She felt warm. The carnival ride feeling was now replaced with an ache in her neck and a throbbing in her head. She opened her eyes and was met with a dull gray sky outside. For a moment she thought it may still be night, but she quickly realized that somewhere behind the rain-filled clouds the sun was already up. She slowly raised her body up, being careful not to move her head too suddenly to avoid any strain in her neck. She gently rubbed the back of her neck before standing up and heading to the refrigerator for the pitcher of water she kept inside. *Hydration*. The thought returned to her mind.

The water was cold. *Too cold,* she thought. But she immediately began to feel a little better as the chill started the slow process of clearing her head. *It was right after someone put on the music and Rafi had asked me to dance,* she thought, trying to piece together the time when she should have stopped consuming the vile margarita concoction. *How many more (drinks or dances) did I have after that? Two . . . three?* She could not recall.

She finished her second glass of water and began pouring a third. She started gulping it down. A feeling of relief was beginning to wash over her when there was a sudden pounding in her head, almost audible. KNOCK, KNOCK, KNOCK. It stopped. She finished the glass and set it down. KNOCK, KNOCK, KNOCK. The pounding started again, this time her mental processes recognizing the sound as coming from outside her head. KNOCK, KNOCK, KNOCK.

"Professor Diaz?" a voice sounded from behind her door. A chill crept up her spine. She recognized the voice as that of a Niklas.

"Uh, I'm sorry. I'm not decent," quickly calling out the first thought that popped into her mind that might delay their entry. *Shit! That isn't going to deter those assholes. Or worse, make them more anxious to get in here.* Eva did not trust the Niklases when Leeds was around. She trusted them much less when they were not on their leashes.

"Get dressed. You have five minutes."

Fuck you, she thought. "I'm not showered yet." She moved towards the ensuite bathroom. It was the only door, other than the front door, that had a lock.

There was a pause. "You have five minutes."

Eva showered quickly and dressed. As she stepped out of her bedroom, still towel drying her hair, she heard a sound from within her apartment. "What the—?" she started, startled to find two Niklases in her living room. One was sitting on the couch, staring at her with a faint smile on his face. The other was standing in the kitchen area, directly across from her open bedroom door, and toward the now opened bathroom door, flipping through one of her cooking magazines. He looked up.

"It's been seven minutes," he said dryly. "We felt it necessary to come in and check on you."

A shutter ran up Eva's spine. The way he said the phrase made Eva wonder if he had been standing outside the bathroom, 'checking' on Eva while she was showering and dressing. A look of revulsion crossed her face. The Niklas on the couch broke into a full grin.

"You didn't come for coffee. And if you came to watch me shower, well, I guess you've had your little thrills, so you can go now," she said, tossing the wet towel on the couch, next to the Niklas.

"Very funny, Professor," the kitchen Niklas said, tossing the cooking magazine down on the counter. He looked past Eva, into the bedroom.

"I do like the view here, though" he said, turning his gaze toward the windows. Both Niklases broke into grins in unison. "But our business is to bring you to Colonel Leeds."

"Well, as much as I'd love to join you both, I've got a few matters here that I need to take care of. But thanks for stopping by," she said. The dull ache in her head was not making her defiance any easier.

The couch Niklas stood up and grabbed her upper arm. "It's an order." Eva jerked her arm from his grip and gave him an angry look. "From Col. Leeds," he continued, his voice softening a little, as he thought about how Leeds seemed to take too special of an interest in Professor Diaz' treatment. He instinctively reached to grab her arm once again, but, changing his mind, simply gestured for her to proceed towards the door. He followed with a not-so-gentle nudge in the back. Eva moved toward the door.

* * *

Eva was alone, and tried to get comfortable on the plastic chair she was seated upon. The chair's design prevented her from achieving her goal. *So,* she thought, *this contour is supposed to match the average American's ass?* She adjusted her position again. *Or maybe an Israeli's?* the thought suddenly leaping into her mind, given her current location. *Who's the genius who thought this was a good idea?* she wondered. Eva was uncomfortable. She had wrongly presumed that she would be meeting Leeds at Tel-Aviv University. Instead, the Niklasas had driven her toward Haifa. The car slowed as it neared the end of Malal Street, approaching the main gate to Technion University. Eva had never been to Technion before , but had met several of the professors at various conferences she had attended while working in Israel. The armed IDF soldiers at the gate peered into the vehicle and had a brief conversation with the Nicklas driving before lifting the gate and waiving them through. *Security seems to be considerably better here than at Tel-Aviv.* Eva thought to herself. The car traveled to the east side of campus, passing a building on the right that announced itself as the Visitors Center before coming to a stop. The one hour drive up the coast had given Eva the opportunity to relax, and she appreciated that whatever her feelings were towards her kidnappers, at least the man behind the abduction ensured that she was provided snacks and drinks for the ride. By the time the car door opened and she was led towards the Biomedical Engineering Building, her head was clearer and her hatred of the Nicklasas had calmed to merely a deep dislike. The three had entered the building in silence and proceeded to an elevator which had no floor buttons to choose a destination. One of the Niklases simply held a fob next to the interior panel reader and the elevator began its smooth descent below ground. It was impossible to determine how fast the elevator was moving and, therefore, impossible to determine how far below ground level they were traveling, but Eva couldn't help but think it was a considerable distance.

Eva became restless and stood up. She walked to the other side of the small table that was occupying the majority of the space in the windowless room. She paid no attention as she passed by the steel door. *No doubt, locked,* she mused, not bothering to even try. The florescent lights above shone brightly, providing no warmth to the sterile environment. She walked back around the table. The door knob let out a muffled click as it slowly opened. She looked to the knob as the door swung open.

Christ, she scolded herself, *there's not even a lock* on *the door.*

She heard the all-too-familiar 'Yes, sir!' from a Niklas who, no doubt, had been standing right outside the door.

"I should have known," she sighed as her previous disappointment at a lost opportunity to escape faded. A faint, sour smell wafted into the room. *The ocean,* Eva thought as Leeds stepped through the doorway. *Or maybe cabbage.* He shut the door behind himself and then turned to Eva.

"I am sorry to have brought you here on such notice, doctor, but it's for your own safety," Leeds began.

"Safety?" Eva interrupted, as she began to pull the chair back out from the table. She looked down at the seat and pushed the chair back. "Col. Leeds, I am perfectly capable of taking care of myself." Eva paused, her head began to throb slightly as the symptoms of her hangover began to revisit. *Why didn't I grab another bottle of water from the car?* she complained to herself.

Leeds smiled at her but remained silent.

Eva let out a loud sigh as she placed her hand on her forehead. "I don't see Tel-Aviv as a dangerous place and I certainly don't appreciate being dragged down here by your lapdogs and locked in a room." She did not sound as angry as she had hoped.

Col. Leeds' raised his eyebrows. "Lapdogs? Now now, doctor, certainly you did not mean that. They would be so disappointed to know that you don't share my fondness for them. They do so much enjoy your company." Eva gave Leeds a disgusted look.

"But if you're finished," Leeds continued, sitting on the edge of the table, "then I must advise you that, in fact, the Iranian government has, only a few moments ago, successfully detonated a nuclear device. You are, I am certain, aware of the recent escalation of animosity between Iran and Israel. What you may not know is that Iran has put their nation on high alert and has demanded that

Israel cede all territories to the Palestinians. They have also demanded the immediate release of all Arab prisoners."

Eva looked at Leeds, the disbelief showing on her face. "They wouldn't. Not with Israel's close ties with America."

"They would. They have even gone so far as to demand that Israel pull all its troops out of Jerusalem by sunset this evening or they will take it by force. Iranian troops are mobilizing on the Golan and both Hezbollah and Hamas have stated that their respective armies are prepared to, and I quote: 'Drive the Jews into the sea.' Daesh and Al-Qaeda in the Sinai have made several cross-border attacks in the last hour and have also mobilized their fighters near the Israeli boarder. Israel is in the process of calling up all active and reserve units for deployment and immediate combat readiness."

For our part, the U.S. forces have re-deployed the Mediterranean Fleet, which is heading towards us even as we speak. The President has declared the entire region a no-fly zone and has already shot down an East Syrian combat jet."

"This is madness!" Eva shouted.

"It is nothing, really," Leeds responded. His voice was calm. His eyes glimmered, showing almost a hint of joy.

Somewhere far above, air raid sirens were preparing to sound their alarm, sending tens of thousands of Israelis hurrying to the nearest shelters.

* * *

The day had started out quietly enough. The dark clouds that had been gathering continued to build. It was just after one o'clock in the afternoon and Effra was struggling to keep her mind focused. A soft drizzle made it difficult to see anything with clarity. She had repositioned herself toward Pier Aleph to observe the docking of the *ben Sabea,* which was a grayer smudge against a gray sea. Tarek and Joshua were out there. *Soon,* Effra thought, *the tedium will be interrupted by their wave. Tarek, such a nice guy. Funny, caring. And Joshua. Ahh, Joshua. Perhaps, some day . . . .* She perked up a little at the thought, trying to shake the boredom from her mind. She was visually following the *ben Sabea* approach when she spotted a small boat speeding towards the security checkpoint. A rush of adrenaline raced through her body. She snapped to action and immediately began to take a position down towards the shore, any thoughts of Joshua quickly vanishing. Several other soldiers joined her and they took positions along the shore. The small craft continued its course towards the *ben Sabea,* showing no signs of either slowing or changing course. A naval patrol boat that was responding to the threat turned on its bullhorn and directed the intruding boat to stop immediately. As the small craft continued on what appeared to be an intercept course, Effra raised her rifle to her shoulder. She zeroed her sights in on the boat's driver. He was a young man with dark hair and sunglasses, like many in the middle east. *Sunglasses? On a day like today? Who wears sunglasses in the rain?* she asked herself as she assessed the situation. The captain was talking to the other men on the boat, smiling. Effra prepared herself to assist in preventing any attack on the freighter and placed her finger on the trigger of her weapon. The naval patrol boat called out a command to cease again, but there was no response. The navel vessel turned on its siren. Effra felt sweat bead up on her brow. She moved the crosshairs, which had been aimed at the captain's laughing countenance, to his chest. She steadied herself. The naval patrol boat fired a warning shot towards the offending watercraft. Her target suddenly moved to the left of her field of view as the captain, apparently responding to the warning shot, suddenly cut its engines. Effra re-centered the crosshairs on the captain, who, laughing only moments earlier, suddenly wore a very serious face. The men on the boat suddenly realized that they were the cause of the commotion and they raised their arms and started shouting in English, "Don't shoot! Don't shoot!" They were three Americans who, like all Americans, appeared to be in their late 20s or early 30s, seemingly goofing around on holiday. They had not realized they were in restricted waters and they had not set their iMes to the Translate setting, so they were not initially aware that they had been directed to turn around by the patrol boat. *What the hell are American's doing here?* Effra wondered. *They rarely travel out of the U.S. Unless they're military* she reminded herself. As the patrol boat approached the small craft Effra returned her finger to the trigger and placed the rifle's sights back on the captain's chest. *Bastards! I should shoot you, Mr. 'I'm an*

American so you can't really harm me. Haha I'm so funny.' Scaring me like that. She tightened the muscle in her finger, slowly pulling the trigger back.

"Effra! No! Don't do it!" Zev shouted at her. Zev, who had taken position next to Effra earlier, had already withdrawn his weapon. "You don't know that they're Americans. And even if they are, you know we are prohibited from shooting Americans."

Zev was right. That the IDF had not determined whether there was, in fact, a brief period of physical pain or whether there would be any long or short-term emotional consequences suffered by the victim was irrelevant. The IDF issued a policy that the intentional shooting of an American meant to serve no purpose other than to present an aggressive action to an ally and determined that such action was not in line with IDF ethical standards. The position was supported by the Israeli government which feared that the shooting of Americans could possibly contribute to a chilling of relations between the two nations, a risk the tiny nation was not willing to undertake.

"It's not fair," she responded, lowering her weapon.

"What's not?" Zev asked, his voice reflecting genuine concern.

"Avi should have gotten the chance." she said as she began heading toward the dock to meet the patrol boat.

The Americans were not military. At least not all of them. "I'm Professor David Fisher. We're scientists working on a joint Israeli-American project out of Tel Aviv University," the 'captain' of the small craft informed Effra.

"What project is that?" Effra inquired.

David paused for a moment. "The Covid-19-B2 immunization project."

"And what exactly is your involvement?" Effra said, prepared to take notes for the file.

"Excuse me, Ma'am," the third man interrupted. He had not spoken a word to this point of the interview.

Effra turned to look at the man. He was clearly a soldier, possessing a rigid formality that is only acquired through years of military involvement. The soldier had dark, hollow eyes. As Effra looked closely into them a chill swept up her spine. She shuddered.

"Sergeant Mueller, U.S. Army, 401st of the 720th MP Battalion. Professor Fisher is not at liberty to discuss the details of his assignment. If you have specific questions, you can submit them through regular cooperative military channels and I'm sure you can get the information you need." He stared hard into Effra's eyes and gave a dry smile.

Do I know you? the notion crossed Effra's mind, but she refrained from asking. *You look so familiar.* She turned her attention back towards Fisher. "Professor Fisher, what is your reason for your presence in the Haifa area, and specifically, Pier Dalet."

"Private Alfandari," Mueller said, glancing at Effra's shoulder insignias. "Our presence in the Pier Dalet area was due to the failure of Professor Fisher to properly assess his location. And his shortcomings as a captain," Mueller added. Fisher looked over at Mueller but said nothing.

"Sgt Mueller," Effra began, "I'm simply gathering-"

"Private Alfandari," Mueller interrupted again. "These men are under the purview of Col. Leeds, Commander of Middle Eastern Operations, United States Army. As such, they have full diplomatic immunity. This interview is over." He stood up and grabbed Fisher's arm, pulling him to rise from his chair. "I hope this does not cause you any inconvenience, but I'm sure your commanding officer will understand."

Effra stood to stop them, briefly laying her hands on her IWI Tavor, but Zev, who had been standing quietly behind her to this point, put his hand gently on her shoulder. She turned to him, but he simply shook his head. The reality of the situation set in. There was little that she and Zev, or even the whole squadron could do, other than to start an international incident and, perhaps, cause harm to themselves. These men were Americans and she had no means of preventing their departure. She let out a sigh.

* * *

The buzzing caused by the vibration of Rafi's phone against the glass-top bedside table was enough to rouse him out of his sleep.

He lay still, pretending not to notice. The buzzing began again and he rolled over, turning his back to the phone.

"That's your phone, not mine" Rachael moaned, elbowing him in the rib.

"Who the hell can be calling now?" He tried to sit up in bed but let his body fall back down as he put his left hand to his forehead. "Oy vey! What did you put in those margaritas?"

As Rafi's phone buzzed for the third time, Rachael's phone quickly followed suit. They both sat up and looked at each other, nervously. Rachael grabbed her phone and looked at the screen, already knowing what she would see. "I'm stationed out of the University. They won't deploy me to Gaza or to the Golan." She knew Rafi would not be so fortunate.

Rafi looked over at her and tried a smile. He had the same feeling in his stomach as just prior to the first time he jumped out of an airplane. The rhetoric coming out of Iran had been increasingly belligerent, but the emergency activation of all IDF Reserves lent an air of foreboding that confirmed his gut feeling that Iran had reached a new level of threat. He hurriedly put on his pants and moved to go, but stopped abruptly. "You be careful," he said as he kissed her on the forehead. "See you later." He turned and left quickly, Rachael watching him, but saying nothing. She rose out of bed and felt the lingering effects of the champagne. She would have time to shower before heading to the University so she stood briefly, staring at her naked reflection in the mirror. She thought about Rafi's hands holding, touching, squeezing her; his lips exploring the nooks and crannies of her body; they way she felt in his arms as they fell asleep. *Hear, O Israel! The Lord our God, the Lord is One.* The prayer jumped into Rachael's mind and just as quickly a chill rose through her body. The image of Col. Leeds flashed through her thoughts. His cold eyes and humorless smile caused her to think that he was somehow behind this. She hurried to the bathroom to wash the thoughts from her mind.

When Rachael arrived at the University it seemed unusually quiet. The labs were silent and other than the presence of several U.S. soldiers stationed regularly down the hall she was alone. As she neared the first one, he stepped forward, putting down the biometric scanner that had been pointed in her direction.

"Ma'am. You are a project scientist. You are to report to Sector Ayin – down the hall and to the left. Please proceed quickly."

"What is this about?" Rachael demanded. She was not familiar with Sector Ayin.

"Ma'am, I can only state that my orders are to move all project members to Sector Ayin. It's my understanding that any questions you may have will be answered once you reach the Sector."

"But- "

The Soldier raised his rifle and pushed Rachael towards the opposite end of the hall with the butt, gently, but firmly and made it clear that any further hesitancy to move where she was being directed would be met with more violent action.

Rachael began to make a movement to object further, but decided against it. As she passed the second soldier she paused. She turned her head back towards the first, but he was too far to see clearly.

"Ma'am. I have to ask you to keep proceeding to Sector Ayin."

"Weren't you . . . ." She started but stopped herself and as she continued down the hall. As she passed the third soldier she kept her head straight but stared at him closely out of the corner of her eye. A chill ran up her spine for the second time that day. *These soldiers look exactly alike!* she thought. A hint of panic arose in the pit of her stomach and she turned the corner and spotted two more soldiers standing alongside the elevator.

"This way Ma'am. Please hurry," one of the soldiers called out, motioning her over. The elevator doors had slid open revealing a lone soldier inside. His hands were placed on his rifle and he wore a half-grin.

A sudden urge to turn around and run swept over Rachael, but she kept her legs moving forward in slow, measured steps.

CHAPTER SIXTEEN

North Korea's rapid takeover of South Korea has shocked the world. More shocking is the total lack of response from the United States of North America

Dr. Ho sat across the table from Colonel Leeds. The President and his Joint Chiefs of Staff had left after Col. Leeds kicked them out. Their cups of cold coffee sat on the table, alongside several ashtrays full of cigarette and cigar butts. One still smoldered. The Colonel was still angry and Dr. Ho knew not to disturb him. It would be best to wait until the Colonel chose to speak first.

Leeds stared hard at Dr. Ho. "No gratitude after having presented you with such a gift as I have. Where are your manners?"

"Colonel, you know that I have the utmost respect for you." Ho bowed his head towards Leeds, staring him in the eye and smiling ever so slightly.

"Of course," Leeds said. He felt a slight tightening of his scrotum. Leeds could not shake the uneasiness he felt around Dr. Ho. He did not fear the man, *per se*, but he held an underlying suspicion that Ho was withholding knowledge, knowledge that would prove necessarily advantageous to Dr. Ho. "I suspect you will want to join in the fun as we march into Beijing with the victorious Army of the Democratic Republic of Reunited Korea. The destruction of the South Korean people was entertaining to be sure, but so short-lived. I do so look forward to the carnage that will rain down upon the Chinese. As I suspect you will too."

"It is my dream before we met, Colonel, to gain vengeance against those who have disrespected me. When do you plan to send your army onto the Mainland? I do wish to be present when Chairman Xi kneels before the invading army."

Dr. Ho stared beyond the Colonel and out the window across from where he was sitting. The sun was shining as it always did. Dr. Ho grew to despise the sun, to despise the green grass and the warm air and gentle breezes that always blew. It reminded him of his past.

* * * *

Sing-Ju was a beautiful woman. She was a hard worker and admired by the young men for miles around. Oftentimes, neighbors would see a strange horse and cart parked in front of her home, and they knew another suitor was seeking her hand in marriage. The village was a traditional one, and marriages were arranged by the parents, but Sing-Ju's father cared deeply for his daughter, and had, quite unexpectedly, and without even his wife's knowledge, promised his daughter that she would not have her marry someone of whom she did not approve. Men journeyed from miles around to try to win her love, but she had already chosen the man she would wed, Jin Chin.

Sing-Ju and Jin had met in the fields when she was sixteen years old. At that time, there were still the communal fields and the villagers all worked side-by-side to further the wealth of the People and the Party. They had traded gazes several times working in the orchard, but did not speak until one day while Jin was working at Shēnghuó, the Giver of Life. This ancient tree was said to be nearly 1100 years old, planted by the village ancestors and it stood nearly 40 feet tall. Jin often picked the fine leaves from Shēnghuó, which were used for making tea for special occasions, such as marriage, birth and death ceremonies. Only he and several other young villagers were adept enough at climbing its trunk to work in it. One day, while, making his way up the branches to harvest from the highest leaves, which were said to be the most sacred, he found himself face-to-face with Sing-Ju.

"Oh. I did not expect to see anyone here," he said, shocked to be so close to her.

She only smiled in return. Jin watched her, unmoving, for several minutes. Her hands gently plucked the buds and second and third leaves from the branches near the very top of Shēnghuó. She moved to some of the lower branches and as she continued she noticed him staring at her.

"What are you looking at?" She asked defensively.

"You are picking the sacred leaves. You— " but she interrupted him.

"What? Because I am a girl I should not be allowed to pick the sacred Shēnghuó leaves? Because I am a girl I should not even be up here?" She was getting increasingly angry.

"No, it's not that," he started. He found that he was at a loss for words. "I cannot be here with you." He moved to start his descent.

"It's true then. You do not want to be harvesting near me. If it's not because I am a girl, then what is it?" she demanded. "Why do you watch me so intently?"

Jin stopped. "I am sorry. I have been watching you harvest Shēnghuó and I realize that everyone, myself included, pick the highest leaves because they are sacred. But you, you do not seek the highest leaves, but the most perfect leaves. Many pick from Shēnghuó for their pride, but you give pride to Shēnghuó. You Honor Shēnghuó by honoring the harvest. Your eye is sharp and your hands are gentle. I am not worthy to harvest with such a person."

Sing-Ju softened immediately. She reached out to touch his hand, but he had already begun to retreat. He looked up at her, not staring directly into her eyes. "I must go." He bowed his head slightly and then continued his descent.

Sing-Ju watched him climb down, nimble and sure-footed. She was sorry she was too quick to temper. It was very unlike her. Several times thereafter, she would try to start a conversation with Jin, but he seemed to be working in trees she was not. Even when they were both harvesting from Shēnghuó they seemed to be at opposite ends. She continued to occasionally throw a smile his way. Oftentimes, he would turn away, but if she continued to watch him, when he turned back he would return the smile. Sing-Ju found herself watching him more often. He was kind to everyone and was always willing to help out others. She noticed that he was a hard worker, too, usually one of the last ones out of the orchard. She found herself working late more often, in the hopes of having the opportunity to talk to him alone.

It was several months later when she decided that today was the day she would succeed. It was dusk and many of the workers were already emptying their baskets. Sing-Ju determined to continue to work until Jin did, and would catch up with him on his way to the harvest shed where the leaves from the day's harvest were deposited. She had just climbed Shēnghuó when she noticed that Jin had already filled his basket and was making his way towards the harvest shed. She was holding a pouch from her waistband which was reserved for the sacred leaves of Shēnghuó and had not yet picked a single leaf, but decided to descend. It was late and she hoped no one would notice her fruitless climb. As she lowered herself down from the lowest branch, she turned suddenly to a sound coming from behind. Even from the harvest shed, Jin heard the loud rustle and sound of a small tree being split. Then he heard the screams. A wild boar had charged through the underbrush and was running wildly among the trees. It was an unusually large beast, weighing nearly 175 kilograms and was mad with hunger. There had been rumor several months ago of a wild beast wondering about after several small lambs had been killed or gone missing in the evenings. The last of the straggling harvesters ran out from the orchard and as the villages gathered themselves up, screams echoed from a short distance away. Without even forethought, Jin made a dash toward the area from where the scream came, Shēnghuó. Running past the rows of trees, he spotted a young girl lying on her stomach. Her basket of leaves was strewn before her and her left foot was held at an odd angle. She had stepped into the burrow of some small creature and struggled to get up, but she was unable. Jin ran straight towards her and without breaking his stride, lifted the young girl up and leapt up the trunk of Shēnghuó. His momentum brought him near the lowest branch, and he slung her body over it and continued his arc towards the ground. The boar stopped momentarily to observe the girl, lying on the branch but out of its reach, and charged after the still running Jin. By now, some of the other villagers had gathered their wits and pursued the animal with spears. By the time they caught up with the beast, it had managed to gore Jin in his left thigh, which was bleeding profusely as a large chunk of skin and muscle clung from the place where it had previously resided. He had blood on his torso and had a gash on his neck. His jaw was black and blue and he had blood dripping from his nose. The animal was preparing to charge again. Jin was leaning against a tree, his weight on his good leg. He was teetering about, struggling to move but unable to control his injured leg. The beast sprinted towards Jin but was slowed by the sudden blow of three spears striking its right side. Two fell almost immediately, the wounds not much deeper than its hide. The third spear however had driven in deeper, just behind the shoulder, causing the animal to turn towards the attackers, but only briefly before it took up its charge toward Jin again. As it lowered its head to deliver its fatal blow to Jin, its right fore foot gave out, causing it to roll over to its right and just barely missing its target. The boar raised itself back to its feet and shook its head a couple times before it turned to attack again. From its left side, four more spears hit the target, one of which sank deep into the beast's left side. It let out a brief squeal with the remaining air in its now-collapsed lungs and slid forward to the earth, nearly making its intended target.

Jin was bed-ridden for the next two months during his recovery. It was three days before he opened his eyes, but upon waking, his spirits remained high, as they had always been, though he was unable to speak. Villagers came to visit him throughout the day and Jin would sit quietly as they told him of the events of the village. Although Jin was generally quiet and kept to himself, he did not mind the visits. Especially those of Sing-Ju. She had been there the day he opened his eyes and continued to come to visit for a little while each day, sitting in the small chair in the corner of the room. Others came often, but no one else visited each day. Jin found it both exhilarating and curious that she chose to do so. He had admired her beauty, but could not forget the day he insulted her on Shēnghuó. Since that time, he had not spoken to her. She would occasionally smile in his direction, but it was certainly never directed at him. When he would notice her smiling, he would turn to confirm that one of her friends was standing behind him, that the smile was meant for another. Oftentimes, he was right, and he would turn back, safe in knowing her gaze was directed at another and return the smile. Had anyone been watching this play out, Jin would certainly have appeared a fool, but he would smile back, nonetheless. She was gentle and beautiful and what was the harm of pretending she was smiling at him and he smiling back? It had brightened his days.

But now she was actually in the same room, actually smiling at him. She would tell him of her harvesting Shēnghuó, of the weather, and of other simple matters. Occasionally, she would tell him of the suitors that frequently arrived at her home, the places from whence they came and the manner of their dress. He could tell that it embarrassed her, but it gave her something to talk about and Jin found the information of other Provinces and their people and customs very interesting.

It was another two weeks until Jin was finally able to speak. When he opened his eyes, Sing-Ju was sitting in her usual place. "Good Morning," she said.

"Good morning," he was able to croak out. The sound of his voice shocked both of them and he immediately reached for the cup of water near his bed. He took a long draught as she smiled, looking on.

"How is she?" he asked immediately.

"How is who?" Sing-Ju asked.

"The girl. How's the girl?" But Sing-Ju continued to look at him, puzzled. "The girl with the hurt foot. The one who I tried to save from the wild boar. How is she? Did she survive?"

Sing-Ju's face immediately dropped Her eyes began to water. "You mean, you don't know? Nobody told you?"

Jin felt a wave of sorrow sweep over him. He knew the news would not be good He looked down as he shook his head.

"It was me, Jin. I was the girl. Didn't you know it was me?"

Jin looked up suddenly. Sing-Ju stood up and limped over to his bed. He had never seen her feet before as she sat in the small chair in the corner. She had a cast on her left foot. Things started to become clear to Jin. Sing-Ju was in the infirmary, too. She was recovering from a broken ankle. That was why she visited every day. They were both patients at the infirmary and circumstances allowed her, actually likely caused her, to visit each day. He was oftentimes the only other patient in the place. How else would she spend her time?

But then she was sitting on his bed, kissing his forehead, his cheeks, his lips. Tears were falling onto his face, his neck. "Jin, you saved my life. And ever since the day we met on Shēnghuó, I have loved you, Jin. You did not care who I was or what I looked like. You saw me for who I am. And when you came to rescue me, you didn't know it was me. You only came to rescue someone in need of help. You are a kind man, Jin, and I love you." She laid her head on his shoulder and continued to cry. He placed his hand on the back of her head and gently stroked her hair. He thought about what she just said and his eyes began to water, too. "I love you, too, Sing-Ju. I love you, too."

Political events in the capital began to interfere with life in the village. Shortly before their engagement, Beijing issued its latest 5-year plan which called for the people to make profits and resulted in a continued shift towards private property and capitalist markets. As a result, many in the village began their own tea tree farms, and they would work both the communal and their private orchards. The profits realized by working for oneself was not lost on the villagers and within a short period, most of the villagers were spending the lion's share of their efforts on their own farms, and contributing a bare minimum to the communal farm. While this worked out well for most, the few villagers who were elderly and had no children or were otherwise disabled and had no means to operate their own private orchard began to suffer. Jin and Sing-Ju led an effort to secure volunteers from every household to devote a portion of their efforts to the communal farm for the benefit of the needy. It took effort and persuasion, but they succeeded in securing promises from all their neighbors.

After they wed, Jin and Sing-Ju continued to devote time to helping their neighbors, and often worked full-days in the communal fields. Even though this would require them to work late into the evenings to tend their own tea tree farm, they both felt charity was a joyful responsibility. More so, however, they cherished the late evenings they would spend in their orchard after the other villagers had long since turned in for the evening. Working under the stars provided the young couple with the freedom to do as they pleased and the privacy to not be disturbed. They worked hard, but managed to find time for themselves. As their own orchard grew and flourished, their love grew, and soon Sing-Ju found herself with child.

The young couple kept the news to themselves until nature prevented their secret from being kept any longer. Sing-Ju continued to work as hard as ever, but neighbors and friends tried to get her to slow down, for the sake of the baby's health, and her own. Nothing anyone could say, however, stopped Sing-Ju from working both the communal and their private orchards, as well as helping tend to the sick and keep a home for her family. The young couple was eager to share their excitement with everyone and many of the villagers often stopped by to check on Sing-Ju.

It was during the 34th week of her pregnancy that she collapsed. She had been working in the communal orchard since daybreak and when Jin came upon her body lying beneath Shēnghuó, he let out a cry that sent most of the village running to the fields. When the first to respond arrived, they found Jin cradling Sing-Ju in his arms, tears trailing down his cheeks, rocking her slowly back and forth. They thought that Sing-Ju was dead, but the village doctor quickly dispelled the error.

"Jin! We must get her to the infirmary immediately!" But Jin was unresponsive. He had gone into shock and as several of the men removed Sing-Ju from his arms, he continued to sit upon his knees, rocking slowly back and forth. Sing-Ju and the baby were the first concern and they were taken immediately to the infirmary. It would be several hours before they were able to persuade Jin to stop struggling against the attempts to put him there, too.

Jin awoke as if from a nightmare. He sat bolt upright, momentarily confused by his unfamiliar surroundings. This was not his bed. Sing-Ju was not by his side. The fogginess subsided quickly. He was lying in the infirmary. It was night and the whole place was quiet. He quickly got out of bed and walked into the room next door. He was both relieved and distressed by what he saw. Sing-Ju was lying still, breathing in short, shallow breaths. His relief came from the fact that had she been dead, the bed in the only other room in the infirmary would have been empty. His heart felt a brief moment of relief before it tightened again. *Yes, she is alive, but is she okay? And the baby?* No answer could be had until morning, so he sat down and waited. And kept watch over the only thing that really mattered to him.

It was over twenty hours before she opened her eyes. Jin was sitting in quiet watch and immediately jumped up. A smile slowly crossed her face and she let out a small laugh, that was quickly followed by her grimacing. She was clearly in some pain.

"Please, do not exert yourself, my love. Lie still, Sing-Ju, rest."

"Am I okay? Is the baby okay?"

"You are okay. Chou says you work too hard. You try to do too much. The baby is okay, too. But you must lie here until he comes. You cannot risk moving around and harming the baby."

"But the orchards. The tea." She attempted to sit up, but quickly stopped when she felt the sharp pains.

"I will be here with you, my love."

"No," she said. "You must tend to the orchard. We have much work to do before the harvest."

"I have been tending the orchard. And you. I will be here when you are awake. I can work while you sleep, at nights."

She smiled again. "Then the favor is returned."

Jin gave her a puzzled look. "You don't remember, but I tended to you, too. After you saved me from the boar, I kept vigil over you, in this very room." She looked around. "It is how we began. It is only right that it is how the baby's life will begin. Our love is tied to this room. He will enter this world, a part of us, a part of our love." She smiled at the thought of the completeness of their lives. Jin forced a smile, too. But until Sing-Ju and the baby left the infirmary, he would not relax.

Jin remained in the infirmary during the day, always available when Sing-Ju would awaken from her many naps. During the night, he would work in the orchards, spending time in both their own orchard, and the communal orchard. At times he felt like collapsing, but seeing Sing-Ju slowly regain her strength gave him energy. Her strength was returning and she had begun to get restless. She wanted to get out of bed, to help Jin in their orchard. They began to speak of the time when she would return, when they could work side-by-side. Sometimes, the conversation would be tinged with a shadow of resentment. Resentment that they could not share this time together, resentment that they would never be able to recapture the lost time together. They spoke often of their love and Jin would repeat that Sing-Ju was the most important person in his life; his reason to be.

In the days leading up to the delivery, Sing-Ju began to experience severe headaches and shortness of breath. Jin would attempt to relax Sing-Ju by reciting poems. He would not allow her to her respond, in fear that it would exacerbate her shortness of breath, and for days, anyone near the infirmary would hear only Jin's gentle voice and the sound of Sing-Ju's short quick breaths. When Sing-Ju's water broke, her breathing regained some normalcy and it was with much excitement that Jin awaited the arrival of his newborn child. The labor was long and Jin was often removed from the infirmary by his good-natured neighbors who knew that the screams of Sing-Ju caused Jin much strife. It was after one of these short respites that Jin returned to the sound of a baby crying. He rushed into the room to find the midwife holding his young son. Sing-Ju was clearly exhausted and out of breathe. Jin smiled at her, knowing her hard work was over and that they would now be able to enjoy their new son. Sing-Ju's mouth began to curl into what Jin thought would be a smile, but it quickly ended and instead, her mouth began to twitch. Jin moved towards Sing-Ju and as he reached down to touch her, she went rigid, her eyes rolling to the back of her head. Suddenly, her whole body began to convulse in erratic movement and Jin noticed blood and foam begin to appear around her lips. The midwife let out a scream and tried to push Jin out of the room as she held the baby near her own body. Sing-Ju suddenly bolted upright, eyes staring out into nothing, and Jin watched in horror as the sheets below her began to turn a shade of crimson. The blood quickly saturated the bed covers and began to pool on the floor beneath her. As quickly as the seizure had started it stopped. Sing-Ju fell back onto the mattress, her skin a deepening shade of blue as the pool below her spread across the floor. Jin had managed to move around the midwife and was kneeling before his dead wife, kissing her cold lips as his tears splashed on her face.

As a young child, Ho was aware that he was looked upon as different for not having a mother. Ho misconstrued his classmates' sympathies for condensation and he became withdrawn and, oftentimes, lashed out at classmates. By the time Ho was twelve, the other children had acquiesced to his apparent desire to be an outcast. They taunted him and called him names, but never in the presence of the boy's father. They respected Jin too much to dishonor the boy in front of the father.

In an effort to combat Ho's penchant for cruelty, some of the children began to spread rumors that Ho was responsible for Sing-Ju's death. Whispers circulated that Ho had willed his mother's death at birth, out of pure spite. Others claimed that as a fetus, Ho heard Jin's confessions of love to Sing-Ju and was jealous of his father's love for Sing-Ju. Others simply believed the baby was possessed of an evil spirit and that it sought only to spread pain and hurt.

Ho was a bright child, but had little aptitude for farming. He spent his days studying books and experimenting with whatever he could lay his hands on. Jin was not troubled by his son's curiosity when he was younger, but as he grew and was expected to take on responsibilities on the farm, Ho proved to be a weak and unmotivated child. Jin would have to hound him from morning until night to perform his chores. It kept Jin from doing his work and Jin soon found it more efficient to simply let the boy wonder and handle all the farm duties himself. It made his hard life more difficult, but Jin was a kindly man and Ho was his son and only living relative. So while Jin was forced to work harder than most, and was often frustrated by the boy, he loved Ho unconditionally and defended him from the evil stares and angry words of their neighbors. Ho, for his part, seemed to either ignore the other villagers, or was completely unaware of their existence. He continued on his studies and experiments as if he were alone in the world – ignoring even the occasional acts of kindness from other children in the village who would ask him to join in on their games every now and then. Sometimes at night, when the sun had vanished from the skies and only the pale glow of the moon and stars lit the night, Jin would find the boy lying in his bed, still and quiet. He would sit beside his son and stroke his hair gently, so as not to disturb his sleep. He would whisper his love to the boy and forgive him his shortcomings and the death of the mother he never met.

The villagers' disrespects weighed heavily on the boy and in the late spring of the year he turned twelve years old he devised a plan to seek his revenge and gain the envy and respect of the others. For reasons unknown, the normal spring rains had suddenly ceased. The mid-April skies had failed to provide the earth her much-needed rain and when May had come and nearly gone, several farmers noted that their tea trees had begun to whither and take on a reddish hue. An unknown malady added to the drought conditions and this blight spread quickly among the tea farms. The dried reddish-brown branches of the rows upon rows of tea trees stretched across the landscape. It was as if the very Earth herself had suffered a mortal wound and they were witnessing a slowly spreading stain of blood. By mid-June, temperatures spiked to the low 90s and remained there well past late summer. The much-needed rains from the monsoons never came as the drought continued and the disease spread, becoming known simply as "The Curse". Each day The Curse edged closer and closer to the Chin family farm. Each day, Jin Chin, Ho's father, would come home and report to his son of the loss of another neighbor's crops to The Curse. It was not long before the Curse spread out and past Ho's farm, on its way to destroy the lives of those in neighboring villages, yet for some unknown reason, the Chin farm remained unscathed. Jin became troubled by this fact. Yet he noticed his son appeared not only unconcerned, but seemed unsympathetic to his fellow villagers. More troublesome to Jin, was the uneasy feeling he had that Ho somehow knew something about The Curse, and why it spared their farm. Ho became boastful of his fortune and, when his father was not around, often reminded his classmates of the largess his father that kept them all from starving to death. Jin's trees continued to yield the increasingly valuable leaves and the Chin family fortunes increased. No longer were the

Chins poor farmers from the outlying Xihu area, but wealthy tea traders able to name their price for their increasingly valuable goods. Jin was, however, a kind and modest man and he was careful to always seek enough for his goods so that he was able to provide not only for his son, but for his neighbors, so that no one in the village was wanting for basic needs, but not so much as to place himself firmly in the middle class. Jin could not explain the blessings that he had been granted by nature but he thanked the ancestors for sparing him The Curse and showed honor and respect by spreading his blessings among his fellow villagers. Jin would deliver goods and supplies to them anonymously, sneaking out at night and leaving packages at the foot of their doors. Jin's acts of stealth were unnecessary. It was clear to all that such gifts could only come from the House of Chin – but his neighbors accepted the gifts of life without complaint or compliment. To burden Jin with knowing others were indebted to his generosity they felt was too heavy a stone to have him carry. Jin was already burdened with a son too lazy to work and a wife who died in childbirth.

It was on a return trip from Lincang when father and son stopped for the night in an area of the Yunnan that still had healthy tea trees when Jin became suspicious. He had awoken suddenly in the night from a dream in which Sing-Ju was crying and unable to be consoled. When Jin had asked what was wrong she shouted a singular word '*Genesis*', which he did not understand. He asked what she had meant and, in asking, awoke. The bed mat beside him was empty. He sat upright, his eyes searching the small room but Ho was not there. He lay back down, fearful of where his son had disappeared to, and repeating the nonsensical word *Genesis* to himself when he heard Ho sneak back into the room, slip his shoes off and lay back down on his sleeping mat. In the morning Jin awoke early and noticed a tea leaf stuck to one of Ho's shoes. Although they were in tea country, there was no point on their journey where they actually walked among the trees and an errant tea leaf simply would not exist in a time of such limited crops. Jin said nothing and they returned home.

It was a week later when Jin had learned that the blight had struck the tea farms near Lincang. Jin could not remain silent any longer.

"Ho. You must tell me where you went that night we spent in Lincang."

Ho spat at his father. "You stupid old man! You are the reason this Village survives, the reason they eat. But they disrespect me, your son. They have always done so. To create such a curse was an easy feat. To see its success a joy."

"What are you saying, son? You . . . you?" he could not comprehend what Ho was inferring.

"It is my revenge, father. *Our* revenge."

"No son. Not my revenge. I do not want such," he sobbed. "Why? Have you not enjoyed our Village? Why such hurt? Why such hate?" The old man broke down in tears.

The father's sobbing had no effect on the son. But as quickly as the tears had begun, Jin suddenly rose up, his face red with anger. "Out!" he shouted. "You have brought shame and dishonor upon our house! You have disrespected your mother's spirit." The old man broke down again and sobbed, turning his back on his son.

"But I am making us rich, Father. We will soon be the most powerful family in the province, and eventually all of China! No longer will they laugh at me, or point their fingers. They will respect me!"

The old man did not turn around. Ho moved to gently place his hand on his father's shoulders. "But Father—" Jin shuddered at his son's touch and he jerked his body away from his son, saying nothing.

Ho's hand slowly returned to his side. "But Father . . . I did this for you. For us. Father . . . please . . . ." His words were met with silence.

Ho left the room but returned a short time later carrying a sack of his belongings. Ho looked to his father, but Jin remained as he had been. Ho let out a sigh and left.

CHAPTER SEVENTEEN

Col. Leeds, Operation Phoenix is scheduled to begin in T-minus Three Minutes. All systems are go.

Eva had been doing very little since she was brought to Technion a little over a week ago. Her questions as to why she was there and what purpose she was serving went unanswered. She became resigned to sitting alone playing solitaire until the she was called to meals. She had only just recently finished lunch and was finding success in her current hand. So much so that she did not pay attention to the recent announcement. *I may actually win this game,* she thought. She had just turned over a red seven when a Niklas grabbed her arm and pulled her out of her chair.

"What the-!"

"Shut up and come with me," the Niklas commanded.

"As if I have a choice," she responded as he pulled them both into an elevator. It started a rapid descent. When it finally stopped, the Niklas pushed her into a room and shut the door behind her. She turned to shout back at the Niklas when she noticed Col. Leeds standing in the room, staring at nothing in particular. Eva turned towards Leeds when the announcement from a hidden speaker in the room updated the status.

"Operation Phoenix is scheduled to begin in T-minus two minutes."

She swallowed hard and struggled to keep her composure.

"What is Operation Phoenix?" she ventured.

"That, my dear, is why I have taken the trouble to bring you here to Technion. We are about to see if our science works. Most exciting, don't you think?"

Eva turned towards Leeds, any thought of keeping her emotions in check long since forgotten. "What are you talking about, Colonel?" her voice raised.

"There is no need to be alarmed, Professor. I am confident this little experiment will turn out as we expect. You see, there is little point in having developed a system to basically recreate individuals at a whim if we cannot be certain it works." Leeds smiled. "As we are speaking, Iran has just launched a battery of missiles, including two nuclear armed Sajjil-2 missiles. One is targeting Tel-Aviv." He paused to see Eva's reaction.

She took a breath as she considered the possibilities. "You dragged me up her to protect me? What about the others? What about—"

Leeds' smile broadened. "Now, now my dear Professor. You need not worry that I have brought you here for your own protection. The second nuclear warhead is aimed right here at Haifa."

Eva's jaw dropped open. No sound came from her parted lips.

"Of course, we are more than safe here. This facility is designed to withstand any threat of nuclear attack." Leeds gestured to the entirety of the room.

Eva let out a short sigh but felt no relief from the uneasy feeling that had been building inside her since the Niklases had knocked on her door that early morning a week or so ago.

"I have brought you here, Doctor, to assist in the little project we have planned. It's all very simple. Intelligence shows that the Iranians will drop two nuclear devices on Israel, instantly killing approximately 2.7 million people. We, of course, have replication data on each and every Israeli and through the hard work of your lab, we can duplicate each and every one to the moment before their death."

"What are you talking about?" Eva interrupted. The Israelis haven't been inoculated. There has been no systematic distribution of PreVentall. How do you expect to have data on these people?"

Leeds spoke knowingly. "That is where you are mistaken. We have inoculated the entire population with a modified PreVentall.

True, the population believed they were receiving only a Covid-19-B2 vaccine, but the oversight will be forgiven - given the alternative." Leeds looked for a reaction from Eva, but saw none. He continued. "The PreVentall inoculation distributed here is capable of mapping out the cellular composition of every Israeli and we have been tracking and storing that information on the entire population."

A puzzled look crossed Eva's face. Leeds smiled at the delayed reaction. He continued. "This particular PreVentall will neither prevent aging nor harm, unless, of course, we choose to allow that, such as now. There will be several million persons vaporized in just a few minutes. And if we have done our science correctly, we will replicate each one and return them to circulation. Using ReGenbots we can reduce or negate the harm to those not killed by the nuclear explosions, and along with nanobots to scrub all traces of radioactive materials from the area, it will be as if nothing more than a large explosion has even happened, thereby negating any evidence of a successful nuclear launch by Iran.

"What I need from you is to assist in assessing the success of the replications. And of course, should something not work as planned, I will need your creativity to help solve the problem. You have full access to the replication lab here. You are free to move about as you wish. However, should we come across any unexpected issues, I will need you to be available to assist in developing a resolution. Failure is not an option."

My god – they planned this – they fucking planned this! The thought raced through her head. She struggled to stop her mind. "Failure?" Eva questioned. "Colonel, we're talking about 2 million innocent lives."

"The lives don't matter, Eva. This is a strategic operation. Think of it. Iran will have successfully launched two nuclear warheads on Israel and the world will see, and Israel will proclaim, only the results of an unsuccessful attack from several conventional warheads. Of course, this is a secret, so there is some necessary collateral damage; some deaths, some damage to property. The Iranians won't know what to think. They will have lost what they believe to be the decisive edge in their military plans. Although the

Israelis also possess nuclear weapons, they would never resort to using them except to avoid complete annihilation. The Iranians have less reservations about that. So by creating this little diversion, we can bring military balance back to the Middle East." Leeds smiled and stared at Eva, who remained expressionless.

"Ah, I do suppose that is of little concern to you. Then think of the people. We must be successful on their account."

Eva turned away. She was filled with a mix of anger, fear, hatred and disgust. She took a deep breath.

The lights in the room dimmed off and on and the blare of an electronic horn sounded. "It would appear that we are under attack," Leeds said calmly.

Eva's Hebrew was minimally proficient but she caught enough of the programmed message to understand that the warning was advising all persons to seek shelter.

* * *

It was not so much the several hours of paperwork that Effra was required to file due to the foolishness of the Americans, but the fact that she would now be late leaving for her afternoon break that angered Effra. And, of course, there would be no greeting today from Tarek and Joshua. Effra was in a foul mood as she submitted her report.

"We just received orders to extend daily duty today. You can take your break, but stay close," the lieutenant who took her report advised. "And make it short, Effra. Something's up, but I don't know yet what it is." It was nearly three o'clock, much later than the first time she had met Noa.

"You know me, lieutenant. I only need about fifteen minutes," she replied. Notwithstanding the days' events and the gloominess of the day, her mood lightened at the thought of seeing Noa and she hoped she would not be too late to at least say hello to her young friend. As Effra raced out the door and up the street, the rain began to fall a little harder. She spotted a blue Lupine flower growing from a crack in the sidewalk. Effra bent down to pick it to give to Noa as an apology for being late. Noa would certainly have been expecting her by this time, and Effra wondered again whether Noa would be worried. She quickened her pace.

As Effra neared the edge of Hazikaron Park, across the street from Safrad Day School, she spotted Noa anxiously scanning the park for her. When she spotted Effra approaching a smile swept across her face. Effra waived excitedly, her arm making a large arc over her head. "Noa!" Noa immediately ran around the school fence and across the streets towards Effra. As if swapping the masks of Thalia and Melpomene, Effra's grin had magically appeared on Noa's face and Effra's face expressed the anxiety that Noa had been wearing only moments before.

Effra raised her hand and shouted, "Stop!" fearful of repeating the events of their first encounter, but her voice was drowned out by the sudden shriek. She quickly glanced to the left and right, searching for the vehicle she imagined was only feet from striking Noa. Noa stopped suddenly and looked up, and Effra quickly followed suit. The fear of Noa being hit that had arisen in Effra was immediately replaced with the realization that threat was not from the streets, but from above. The whine of air raid sirens continued their slow, pulsing warnings. Effra looked back towards Noa. Already, people had started running towards the closest bomb shelters. Cars were stopped in the middle of the street and people were jumping out, following those who were already disappearing below street level. Noa, however, seemed to ignore the commotion and continued to run towards Effra. Effra tried to keep one eye on Noa as she again looked quickly around. She was not familiar enough with the area to know the location of the closest bomb shelter. As Noa continued to near, Effra spotted the other school children running towards the far side of the school. Closer and to the left, a stream of Israelis were descending a steep staircase. Effra ran towards Noa and swept her up as the met. She immediately began to run.

"Oh, Effra," the little girl began to sob. "I thought you weren't coming." She nuzzled her head into Effra's shoulder.

"I will never stop coming to see you, my little Noa," Effra replied. She, too, started to cry. The shelter suddenly seemed so far away and her legs felt as if they were moving in slow motion. At the same time, she was overcome with the feeling that something was approaching her from behind. She held on to Noa more tightly and tried to run faster. She felt the little girl's head lift, her warm breath rushing softly against her ear.

"Look. The sun is coming out. You and the sun both came," Noa whispered as she nestled her head into Effra's neck.

Everything in front of Effra suddenly turned unnaturally bright, illuminated by a light source coming from behind. At almost the same time she felt a sudden rush of the hot breeze overtake her. She tried to scream as the wind on her back intensified but she seemed unable to catch her breath. There seemed to be no air whatsoever. The young girl was still nuzzling Effra's neck when she shifted her position and lifted her arm from around Effra's neck. "Look, I can see my bones . . ." Effra stopped running, unable to move any further and squeezed Noa tighter, trying to protect the young girl from whatever horror was happening around them. The hot breeze at her back became a ferocious wind that lifted Effra off her feet. She spotted a pink bunny floating in the air in front of her before everything went suddenly black.

CHAPTER EIGHTEEN

Reports that Iran has launched a nuclear attack on Israel are unconfirmed. The Iranian State Radio has announced that along with a battery of long range missiles, it fired two nuclear warheads at Israel and that they successfully hit both Tel Aviv, and Haifa. Iranian news sources allege that based upon targets, casualty levels should be excessive. The Iranian President has announced that today will be a national holiday and proof of the power of Allah. Israeli Defense Forces spokesperson, Elat Benowitz has stated that multiple long range missiles from Iran were directed into Israeli airspace. Most were intercepted and destroyed prior to reaching their targets, however, several missiles did penetrate defenses in Tel Aviv, Haifa and Jerusalem. Israeli sources indicate that casualties are expected to be minimal, with the largest number of possible deaths occurring in Jerusalem. No further information is currently available . . . .

When Clay awoke, Lillian was gone. He looked out the window. '*It is 11:27 a.m.*' he was advised. He grabbed his pants and pulled them on. He knew that no further reports would be transmitted. Americans liked to be mildly informed, but quickly moved on to other or new matters without thought of the past. He suspected that by two o'clock most people would have forgotten the matter entirely. Back to their lives of ease. Clay went to his laptop to see if he could find any additional information on any foreign news sites. Sites that would not show up on any internet search, but were available to those who knew the url.

Israeli government officials have confirmed that a Jerusalem-bound missile from Iran struck and destroyed the Al-Aqsa Mosque. Over two hundred and fifty Muslim faithful were killed by the explosion and over three hundred injured, many in serious or critical condition. In addition, nine Israelis and three foreigners praying at the Western Wall were injured by debris, but all have been released from local area hospitals. Iranian officials claim that Israeli and Western sources are providing false information and that its sources have confirmed that two nuclear missiles impacted Israeli territory. Iranian news sources are quoting government officials that the death toll in Israel exceeds two million victims and there is widespread destruction across the nation.

Clay typed in the url to the British based *Al Jazeera* news site. He scanned the pages for more information. Reports coming out of Jordan indicated that six very large explosions occurred over Israel and eyewitnesses spoke of observing several large mushroom clouds and experiencing what can only be described as a blast wave followed by temporary impaired vision. These claims, however, were dismissed by Israeli officials and live pictures coming out of Israel showed no signs of any significant damage. The Israeli officials described how the explosions resulting from the interception and destruction of the five successfully intercepted missiles could be confused for mushroom clouds and note that all claims of injury have proven to be minor. Additional video showed the exploded remains of several Iranian Maleki-Moat-3 missiles and the massive destruction at the Al-Aqsa Mosque, with dozens of IDF forces lending a hand in clearing debris and searching for survivors. Clay watched a video review of the destruction. Several pairs of IDF soldiers moved in unison through the streets, calling out orders to each other, moving materials and victims. They appeared calm and in control while dozens of Arabs were screaming. The footage broke to another scene of ambulances and rescue workers running through the streets. Clay noticed in the background an old Arab being turned away by a soldier and moving to the next soldier, apparently hoping for a different response. The old man suddenly stopped talking and leaned in close to the soldier. He appeared to stumble back and then ran off with a dexterity that surprised Clay. Clay stopped to take a closer look but the image changed and the scene was gone.

The news continued to be updated. Hamas leadership had made threatening remarks towards Israel, accusing them of purposely failing to intercept the missile that hit the Al-Aqsa

Mosque but were not speaking about the obvious fact that Iran had launched a missile directed towards the holy sight. The Arab League was quick to denounce the Iranian attack and the destruction of the Al-Aqsa Mosque and Sunnis from across the region were seeking revenge against Shi'a Muslims. A spokesman for the Secretary of State, flanked by two soldiers in dress uniform, warned Iran against further attacks and stated that the United States was leaving all options on the table to address the act of war against Israel. Although the two soldiers were slightly out of focus, Clay recognized them as Nicklases. It suddenly struck him that several of the IDF soldiers shown taking defensive positions for whatever may follow had a strong resemblance to the Niklases. It is likely what surprised the old Arab – a group of Israeli Nickalses. "Jesus . . ." he turned away from the screen and shut the laptop lid. Silence.

What the hell is going on? Clay wondered. Iran admitted to destroying Al-Aqsa Mosque in its attempt to prove it launched a successful attack on Israel, but there was absolutely no evidence whatsoever of any real damage other than to Al-Aqsa. Iran was adamant about firing two nuclear equipped ICBMs and some neighboring Arabs were talking about mushroom clouds. None of it seemed to make sense. "Call Eva," he directed.

"There is no response. Would you like to leave a message?" his iMeme asked.

"No." *Damn! Eva would have more information,* he was certain. He considered going over to her apartment to see if he could figure anything out, find any clues. He still possessed the key she had given him to check in now and then and make sure everything was alright. He knew she was out of the country. Miguel and Jennifer, Eva's parents, were at their second home in Phoenix and he had been consistent in heading to her place to water her plants and keep an eye on the place.

Clay almost immediately dismissed the idea of going there. He did not desire to go digging through her personal items like some type of thief. He thought for a moment. The idea struck him from a lost corner of his mind. The ham radio. *There's got to be a few ham operators out there,* he reasoned. He ran down to his study and flipped the power switch on the radio. The radio hummed to life and he slowly began to scan the frequencies. *Wrong*, he thought. *Let's be methodical here.* He moved the selector to Band 2 and spun the tuning knob counterclockwise to start his search at 1.6 MHz. The information printed on the radio indicated that Band 2 would likely yield only local information, moving from POLICE on the lower frequency past various GOVERNMENT designations in the middle-range and to AVIATION on the far right. As he expected there was no activity. As he switched to Band 3 and swept the tuner back to the far left to 4.5 Mhz. He recalled when he was a kid driving across the country with his family. The family car had an old AM analog tuner. As you turned the dial an orange bar would shift left or right to let you know what frequency you were honing in on. He remembered oftentimes continuing to keep turning the knob long after the orange bar had reached its limit, the wire controlling its movement slipping on some pulley so that you could not break it, but the orange tuning bar indicator oscillating ever so slightly forward with each turn, seemingly struggling to break free of its imprisonment. Back then the radio waves were filled with music and talk and news and commercials. At night, the AM stations would sometimes carry for hundreds of miles and you could listen to radio stations in cities far from where you were, or where you were heading. After the introduction of digital music players, radio stations continued to survive. Markets shrank and programming was increasingly automated, but it remained a viable source of entertainment. It was around the time the cloud became common that radio stations started disappearing altogether. One could listen to the music of their choice on their cell phones without commercial interruption. Some deejays remained popular and broadcast their radio shows via podcast, but as stations disappeared and podcasts became part of the norm, music and entertainment moved from the AM and FM frequencies to the microwave and into the infrared. It struck Clay that he had no idea how information was transmitted any longer. Were they still using microwave and infrared, or perhaps technology and demand had driven communication into the ultraviolet and, maybe even x-ray or gamma ray range?

As he slowly moved the dial clockwise he was confronted with volumes of dead air. He pulled the microphone close and pressed the button: "Hello . . . ." He pulled open a drawer to look for his radio license call numbers. His license had long since expired, but it briefly struck him that he might need to provide some type of identification. The thought quickly expired as he was greeted with more silence. "Hello?" he tried again. As he was spinning through the frequencies he stumbled across some static. He slowed down and backed up, seeking some type of signal beyond the static. He rotated the bandspread knob. Nothing. He continued up the dial. "Hello. This is the United States." Clay smiled, recalling Pink Floyd's *Young Lust* and wondering if anyone was destined to hang up on him. He continued slowing moving the dial. As swept past 7 MHz he heard voices conversing. "Hello?" The conversation broke.

"United States?" in heavily accented English.

"Yes," Clay responded. He heard a brief statement in an unrecognizable language, but it sounded German to him.

"Yofi! This is Israel. You will see. America, you have heard, no?"

"He has not, certainly! Listen Joseph, you must-" said a second voice breaking in over the radio.

"No, David! You must listen. I am not mad," here the voice broke off and Clay heard him draw a long breath, labored and broken, as if he were ready to cry. "David," he said much softer now.

"Joseph Alfandari, my brother, do you know what you are speaking?" David said, toning down his voice in response.

The voice of Joseph Alfandari continued. "Iran has dropped the bomb on Israel! Actually, dropped several bombs on Israel." Joseph paused briefly and Clay could hear him sob slightly. "The destruction was horrible! I have never been so afraid in all my life."

"Yes," Clay responded to the original query, softly, unsure if the answer even mattered. "I've heard reports that Iran launched many missiles and two nuclear bombs towards Israel and all but a few were successfully intercepted and destroyed."

"Lo, lo, lo, lo," the voice had slipped into its native Hebrew. "No! You did not listen. Iran dropped *The Bomb* on Israel. A nuclear bomb –"

"What?! Nuclear bombs?" Clay interrupted. "No, it can't be. I've seen pictures coming over the cloud showing minimal damage –"

"Listen. It is true. I cannot tell you how or why. But I saw the mushroom clouds with my own two eyes. I saw a blast wave. Only now are my eyes returning to normal. The sky turned as black as night. I ran in fear. I tried to contact my daughter who is stationed in Haifa immediately afterward, but no communication worked. I do not understand."

"I tell you, Joseph," the other voice interrupted. "No one knows what happen. Mr. America, you not hear about nuclear explosions? He not hear, Joseph," the voice continued, not waiting for response from Clay. "I am Joseph's brother. I live in Argentina. We hear of bombing of Israel. We see pictures on news. My brother say nuclear explosion, but no explosion on news. My brother not making sense. Tell him, Mr. America. Tell him no nuclear explosion."

"But it not make sense," the original voice resumed. "Effra still not answering. Effra would call me. It not make sense. I ask the world, what happened? I get on the radio and ask what happened?"

Clay could hear the desperation in the man's voice. He suddenly felt like an intruder, invading a personal moment between two brothers, one possibly mad and the other helpless to provide the support and comfort his brother needed.

CHAPTER NINETEEN

All Egged bus lines are free today. All busses will make stops at Carmel Beach Station. Special routes have been added. Please proceed to the main entrance lobby to locate the next available bus. The next available bus leaves for Carmel Beach Station in three minutes.
All Egged bus lines are free today. All busses will make stops at Carmel Beach Station. Special routes have been added.. . . . .

Effra awoke in the sterile room of a medical facility. Her bed was raised slightly and a clock on the opposite wall notified Effra that it was 8:27 but she did not know if that was a.m. or p.m. or even what day it was. She lifted her head up to get a better look around and was suddenly struck with the realization that she felt no pain. She wiggled her fingers and toes to make sure that she was all there. Nothing appeared to be missing. It struck her as odd that she would be dressed in her military fatigues in heaven but before she could complete her train of thought, she noticed a fair-skinned woman sitting on a stool at her bedside, tending intently to some equipment. The woman was dressed completely in white and said nothing. Effra felt that to speak or move further would disturb the woman, so she rested her head back on the pillow.

A door to the room opened and Effra moved her eyes to observe a dark-haired women holding a tablet close to her chest entering the room. The dark-haired women came up behind the woman in white, but neither said a word. Effra was not sure if the woman in white even noticed the other woman standing behind her. They both seemed intent on the information flashing across a screen on a piece of equipment just beyond Effra's line of sight. Effra lifted her head again to try to get a view of what the women were engaged with.

The dark-haired woman turned toward Effra now and spoke. "How are you feeling?"

"Angleet? Damn!"

"חשבתי שנדבר עברית בשמים" Effra said softly to herself in Hebrew.

"I'm sorry?" Eva said.

"I said, 'I thought that we'd be speaking Hebrew in Heaven.'" Effra responded. In addition to the mandatory English coursework in school, her mother spoke in English as often as she did in her broken Hebrew at home, and Effra spoke English like a native speaker.

Eva let out a weary laugh. "You're not in heaven. Although there was a time when I thought that heaven must be a lot like Israel."

Effra was suddenly struck by the ridiculousness of the thought that she had died and gone to heaven. "I'm sorry. I . . . I . . . well, I guess I'm not quite sure where I am," she replied.

"I'm sorry", Eva continued. "I thought that you and the nurse may have already spoken. My apologies."

Eva looked at the nurse, who smiled sheepishly. "She seems to be fine," the nurse spoke softly as she stood up to leave the room. "You're in a medical wing of Haifa University," Eva said, directing her attention back to Effra. You were a bit shaken after the explosion and we wanted to make sure you were alright. Do you feel any pain?"

Effra thought that Eva looked tired. "No, I'm fine. I – " Effra stopped in mid-sentence. "Where's Noa?" she suddenly demanded attempting to lift herself out of the bed.

"Who?"

"Noa. I was holding a little girl when the explosion hit me. Where's Noa?"

"I'm afraid I don't know what you're talking about. Who is Noa?" Eva made several swipes on the tablet she was holding. "She's not your child, is she?"

"No, no. She's . . . it's a long story, but I was holding her, bringing her to the bomb shelter to try to protect her. What the hell happened to her?!" Effra's voice was now raised. The door swung open and a soldier poked his head inside the room. "Is there any problem here?" he asked Eva. A second soldier appeared alongside the first the two of them completely filling the doorway.

Eva turned to the soldiers and raised her hand. "It's okay. There's no problem here. You can wait out in the hall, please." The soldiers looked around the room and then at each other. Seeing no immediate threat, they stepped back out of the doorway and let the door shut softly behind them. Effra stared at the now closed door. *Do I know that soldier?* She thought hard and it struck her that both soldiers looked exactly alike.

"I'll see what I can find out for you" Eva offered. "What is Noa's last name?"

It struck Effra that she had no idea who this little girl really was. "I . . . I don't know. She attended the school near the intersection of Hassan Shuqri and Berwald Street. We met only a few months ago. I go over to the schoolyard many days and eat my lunch nearby. She tells me stories of her friends and what she did in class and . . . she's . . . ." Effra stopped. She was not sure what to say.

"You've been through a traumatic experience today. Why don't you just sit back and hopefully I can help you understand. My name is Eva Diaz. I am an American doing some collaborative research here at the University. I do biomedical research and after the recent events, I was asked to help. I'm just going to ask you a few questions and make sure you're okay and, if you have any questions to ask me, I'll do my best to answer them." Eva set her tablet down on a table near the bed and moved beside Effra. Eva placed her hands gently on Effra's arm and raised it. "Do you feel any pain when I do this?" Effra shook her head. "How about now?" Eva performed a number of flexion and extension exercises on Effra, always stopping to ask if she felt any pain. After a few minutes, Eva stopped and sat down in the chair next to Effra's bed where the pale nurse had been seated. "Do you mind if I ask you a few questions?"

"Perhaps."

"Do you know your name?"

"Of course I do! I'm Effra Alfandari, Samal, Caracal Battallion."

"And your family? Who are your parents?"

"Wait a minute! I don't even know who you are or what's going on here. I'm not about to – " the volume of Effra's voice beginning to rise.

"Calm down, Ms. Alfandari. I'm not trying to upset you here, I'm just trying to make sure you're memory is okay."

Effra was breathing heavily, her body tense. The woman beside her remained calm.

Eva sat back and smiled at Effra before she continued. "Can you tell me the last thing you remember?"

Effra's face blushed at her foolishness. Clearly this woman was a doctor of some sort and was trying to help her. She looked over Eva carefully. An attractive woman, likely in her early 30s. Hair pulled back tight in ponytail. White lab coat with a stethoscope tucked in one pocket. The breast pocket had a name plate attached that read DR. DIAZ and contained a couple pens, a small electronic device and what appeared to be a pair of sunglasses.

That's odd, Effra thought. Dr. Diaz sat quietly while Effra continued her inventory. *She is not threatening. And I am certain that if it comes down to it, I can take this Dr. Diaz down with little effort.* She looked toward the door. *I will worry about those two later.* Effra turned back towards Eva and looked into her eyes. *Smart. Confident. This woman knows what she is doing. But tired. She is working too hard. She's no threat to me. Or probably anyone.*

"I'm sorry. I remember walking over to the school yard, actually running because I was late, but I was going to meet Noa, like I always do. Noa saw me and started running towards me when the air raid sirens went off. I grabbed her and was running to the nearest bomb shelter when I saw a flash. Then it felt like all the air was sucked out of my lungs and my back felt like it was fire. Noa was in my arms. I was running and . . . and that's the last I remember. What happened?"

"There are a lot of rumors going on out there," Eva explained. "What happened is that Iran launched an attack on Israel. Many of those missiles, including one directed here at Haifa, were intercepted by the Iron Dome defense system. It appears that your government had recently deployed new, high power missiles in the Iron Dome system. When these new missiles struck their targets, they acted like a flash-bang. They sent out a large flash of light and sound, as well as packing enough power to knock out anyone in the area. Many people believed there had been a nuclear explosion as a result of the combination of effects. Literally dozens of people were rendered unconscious. There were some bombs that did hit their targets, but it does not appear that you were a victim of one of those. You seem to be in good shape. Like others near these flash-bang type of explosion, you suffered from only mild injuries: some bumps and bruises, perhaps a minor burn and, of course, having lost consciousness."

"You said not all the bombs were intercepted. Was there any significant damage? Any loss of life?" Effra asked, fear in her voice.

"Most fell harmlessly. One missile, however, avoided interception and struck the Dome of the Rock."

"The Temple Mount was hit?!"

"Yes. There was significant damage, and high Arab casualties. But certainly no nuclear event."

"My God." Effra turned away, but quickly sat up and kicked her legs over the side of the bed and returned her gaze to Eva. "I've got to find out what happened to Noa. You're certain that I was not brought in with anyone else?"

"I don't know." Eva picked up the tablet again and tapped on the screen. "The rescue crews brought in hundreds of people. We're checking everyone out for injuries and releasing them, but I don't see any record of you coming in with any child." She saw Effra's face drop. "But that doesn't mean she wasn't here. There are so many people, and with as busy as folks have been, I know that the data being entered is not always current." She forced a smile. "You know, folks are more worried about helping the injured than the clerical stuff." She did not recognize any change in Effra's countenance. She tried to sound hopeful, "I'm sure she is okay."

Eva paused a moment. *What the fuck Leeds? Don't you dare tell me we lost a little girl today.* An anger started to rise up through her. *I don't know how someone, especially a little girl, could have been missed during the Covid-19-B2 inoculations - but if you killed an innocent little girl . . .* She realized that the anger must be evident as Effra's eyes began to widen as a look of uncertainly crossed her face. Eva took a deep breath.

"I'm sorry. I've got this . . . er, splitting headache." She forced the smile back to her face. "You know, perhaps she wasn't injured enough to be brought here. Maybe she was treated on site and is safely back home with her family," Eva lied.

"She's an orphan. Am I free to go?" Effra asked.

"Do you remember what you had for breakfast this morning?"

"Yogurt and fresh bananas."

"Lunch?"

"I told you, I was on my way to meet Noa when the attack happened."

Eva smiled at Effra. "Do you remember your birthday?"

"March 14," she responded, curtly.

"I understand your impatience, Samal Alfandari, but I can't let you leave if you're not well. Just one more question." Effra stared at Eva. "Do you remember anything after the flash and the heat on your back?"

Effra shook her head, the added, "A pink bunny floating in front of me."

Eva's sat back, her mouth slightly agape. "Excuse me?"

"Never mind," Effra said quickly. "Can I ask you a question?" she ventured.

"Certainly," Eva replied.

Effra opened her mouth as if to begin, but then held her tongue. She looked Eva up and down one last time and her eyes fixed on the sunglasses in her breast pocket. "Is it daytime or nighttime?"

Eva glanced down at the sunglasses and semi-consciously tried to tuck them further into her pocket. They were issued to Eva when she arrived in Israel. The team was unsure of Israel's technological capabilities. The glasses were specially made to prevent any Iris Scanning cameras from being able to read properly. She had been asked to wear them anytime she was out of the labs. Apparently, the cooperation with Israel had its limits. DARPA did not want anyone, perhaps including the Israeli government, from being able to track their scientists. "Yes, it's still daytime."

Effra stood up and moved towards the door. Eva quickly rose and was straightening out her lab coat when Effra stopped and turned back. "Please keep an eye out for Noa. I want to be sure she's okay."

"I will. I have your information here on the chart. I will be sure to let you know if I find her." Eva turned back to Effra and added, "And be careful, there's a lot of confusion and fear out there."

"Thanks," Effra said genuinely, and exited the room. Walking down the hall, Effra noticed other people coming from the various rooms and all, like her, being directed by armed military personnel towards the exit. She looked at several others, seeing if any of them appeared to notice that none of the soldiers were IDF, they were all American, and all identical. Several other persons wearing white lab coats and dark sunglasses were moving hurriedly from room to room. When she got to the exit, she stopped to assess the situation. She needed to find someone who could help her track down Noa, someone in charge. She decided she needed to find someone who could answer her questions. *Someone has to be in charge,* she thought. *Someone has to have access to information on everyone who entered the University, including Noa.* She walked around the campus but each time she attempted to enter she was denied entry.

"I'm sorry Ma'am. No one is allowed to enter this facility at this time," one of the two identical American soldiers standing outside the doors said. Effra looked at his name label. MUELLER the tag read.

"Sorry," she responded. She quickly turned in the direction she was pointed, but as soon as she reached the edge of the building she slipped around the corner to seek another building. At the next building she approached she was again denied entrance.

"I'm sorry Ma'am. No one is allowed to enter this facility at this time" one of the two identical American soldiers standing outside the doors said. He gave her a steady look. Effra looked at his name label. MUELLER the tag read. She looked back up toward his face. He leaned in toward her and added, "Please don't try again." A chill ran up her back and she quickly turned to move in the direction Mueller had directed. *Perhaps Dr. Diaz can help?* She dared not approach another American soldier. She started asking others in the crowd if they knew who may be responsible, but, like her, they had been recently released from a lab room after being checked out and were trying to find their way back home. It seemed to Effra that half the country was on the campus. She sat down on the curb to think but it was difficult among the crowds. She looked at her watch. She had been wandering around for almost an hour. Giving up hope of finding Noa for the time being, she followed the crowds to the front of the campus where lines of busses stood to carry passengers back to their destinations throughout the area. She decided to go back home and try to find out what had happened to Noa. She climbed aboard a bus and sat down at a window seat, studying the situation. She pulled out her phone to call her father as she watched hundreds of people continue to file out of the University. She did not think to wonder why no one was being brought in.

Inside, Eva stood in the laboratory, alone at last, her head swimming from the recent events. There were no more people to question, no more Israelis left to replicate. The week had started off to be so promising. She should have known that it was only the calm before the storm. She needed a drink. Badly. *How did this all go so wrong?* she thought.

Although she did not want to admit it, Leeds' experiment seemed to be working fine. Eva had, unwittingly been an important cog in the whole project and the past many hours had been very difficult. She was exhausted. And angry. *How many times am I going to let Leeds use me to accomplish whatever the hell he wants? Why do I keep thinking that what I'm doing is good, or will help people - other than Leeds?* She put her head in her hands and started to sob. *God, I'm such a fool. How easy is it to figure out what is happening?* She was not comforted in the least by the thought that hindsight is twenty-twenty. And yet, she was somewhat in awe of what just took place. *What are the logistics of replicating over a million people? What if it didn't work?* The thought put a shudder through her body.

Eva let out a sigh, that ended in a quiet scream. *My god. What if . . . ?* A terrible thought crossed her mind and she felt her stomach turn. *What if success was worse than failure? Of course, the death of a million people - people who have, time and time again, been the victims of genocidal attempts - would have been horrific. But it would have stopped the horror, even if only temporarily, that certainly lies just around the next corner. How easy it is to find the utility in human regeneration and replication. After all, why should we suffer if we have the ability to alleviate pain and sickness? Why should we lose our fathers, our brothers, our sons, in wars that are usually started by those who have no personal stake in the game? Men whose sons will not enter combat: men who risk very little, with nothing to lose and vast riches to gain. It is no secret that those whose livelihood is derived from war enjoy great benefits regardless of the actual outcome of battle. Government demands no promises and manufactures provide no guarantee that their capital contributions will provide success. The ammunitions manufacturer sells his wares, and government pays the bills, win or lose.*

"Once again, Professor, your abilities have provided the mission with success."

Leeds' voice startled Eva and she sat upright, turning from the direction of the voice to hide her bloodshot eyes.

"You used me," she said, gathering her strength as she turned towards Leeds.

"My dear, no one has been used. We have prevented tragedy from befalling on the Israeli population from those Persian dogs. And we have shaken up the whole Iranian hierarchy, the whole Arab world. The Ayatollah will soon be revealed to be the sole reason for the destruction of Dome of the Rock, and he knows it. There will be a lot of misinformation coming out of the Arab media, but I can assure you, my people are at the very moment holding discussions with Mr. Khamenei and he is seeking a way to save face, and his life, in exchange for full recognition of the right of Israel to exist and an open condemnation of Hamas and Hezbollah. So you see, you have been an invaluable cog in bringing peace to the Middle East." Leeds paused momentarily. "You have accomplished what others have failed for decades."

Eva caught the scent of pickled cabbage. It struck a chord in her memory but she was unable to place it. "So you're telling me our work here was a peace mission? Are we prepared to pack our bags and go home now? Mission accomplished?" the phrase jumped into her mind from her middle school history class. A phrase used by President George W. Bush, and later by President Donald Trump, to express the end of threat and conflicts in Iraq and against the Syrian people, respectively. It was true in neither case.

"Eva," Leeds said as he shook his head. He stood next to her and placed his hand on her shoulder. "You should be proud of what was accomplished here today." He gave her shoulder a gentle squeeze before turning aside. "I am afraid I have several matters I must attend to. Try to get some rest, Professor," he said, reaching into his pocket. He withdrew a card key and handed it to Eva. "Room 626 at the Dan Carmel Hotel. It should be up to your standards. The chef is highly regarded and you will find the spa most relaxing. Nikita has the hands of an angel." A faraway looked entered Leeds' eyes and a slight smile crossed his face. "She can massage the knot out of a Lebanese Cyprus. Do look her up.

"I have arranged it so that you will not be disturbed for a few days. You are not expected at the lab and no one will be seeking your company. Take the time to reset." Leeds moved toward the door and as he approached it opened. A Niklas stood outside, his hand still on the door knob. "You have done good, Eva. Do not forget that."

The door shut behind him and Eva was alone once again. *It's a lie,* she thought. *Removing the ability of an enemy to harm your soldiers does not prevent war. It creates war. Why wouldn't we start a war? What hazard firing the first shot when we know our soldiers are safe from harm? And what enemy, no matter how overpowered, is willing to simply roll over? No,* she thought, *I have not helped stop war. I have ensured that it will remain one of mankind's unique traits.* The image of a pink bunny floating in space flashed in her mind. And of a missing little girl screaming.

CHAPTER TWENTY

In the Middle East overnight, all activity in Israel has returned to relative normalcy. Minor injuries continue to be treated and several persons remain missing after the massive panic wave that spread across that nation. Ayatollah Mojtaba Khamenei has ceased his earlier claims of nuclear attack and has been working tirelessly to avoid any blame for the destruction of the Al-Aqsa Mosque. Islamic belief holds that Allah will prevent damage to the Mosque and Iran is now alleging that Israeli terrorism is responsible for the destruction. Shiites across the region are demanding the destruction of Israel, while Sunni factions are calling for revenge against Israel and Shi'a Muslims. Military troops in Egypt continue to deploy to the Sinai. Russia has pledged its support to Iran and has advised all nations to allow the dispute to play itself out without outside interference. China has indicated that it, too, is willing to support Iran in any confrontation with Israel. NATO forces have been placed on alert and Turkey has warned Iran to refrain from any further military mobilization and to refrain from inciting violence. While Turkey remains technically at war with Israel, it is concerned that any armed conflict may spill over its border with Iran and further embolden Kurdish separatists.

The gentle buzz of Eva's phone broke the silence. She felt too tired to even retrieve the device from her coat pocket. It began its cycle of silence to buzz and back to silence. She wondered how long until the call would transfer to voicemail, but as time passed it seemed that it would be easier to answer than to wait it out.

"Eriv Tov," she answered, wondering if it was still evening.

"Eva? It's Clay. Thank God. Are you okay?"

"Clay . . ," she paused momentarily. "What . . . are . . . well," she fumbled for words and finally, "How are you?" She did not expect to be hearing from him. A million evil thoughts suddenly flashed through her mind. "Is everything alright with my folks? What's going on?"

"Is everything all right? What the . . . ? Jesus, are *you* okay?" Clay responded with a nervous chuckle.

"I'm fine," Eva snapped back. "Are my folks okay?" Eva suddenly recognized the situation for what it was. "I'm sorry to snap at you Clay," she continued, her voice more calm. "It's just that . . . well, I certainly wasn't expecting to hear from you and with everything that I've just been through . . . well, I guess I just thought the worst."

"The worst? I was worried about you," Clay countered. He breathed a sigh of relief. "I don't know what's going on, but the news I'm hearing from Israel . . . well, I just wanted to make sure you're okay. And see what really is going on out there," he added. "I was just -"

Eva cut him off. "Listen to me." She looked around the room to make sure no Niklas had entered. She feared one likely still stood just outside the door. "We can't talk right now," she said, lowering her voice. "Call me back in three hours. From my apartment," she added. Her mind began to race. "Wear some old clothes, something without an iMe." She disconnected the call and gathered herself. She had a lot to do. She sent a text to her parents letting them know she was okay and that she loved them both. Her mind returned to the matter at hand. *I should have told Clay four hours,* she scolded herself. *Okay, let's act normal.* She exited the room, and to her surprise, no Niklas was in sight. *Let's just hope Leeds called his dogs off for real.* She was not confident that the next few days would truly be hers, but she did not have time to worry about that now.

Eva crossed the lobby of the Dan Haifa Hotel and stepped up to the front desk. "I'm staying in Room 626. I'd like not to be disturbed during my stay. No phone calls. And no housekeeping, please. I have some important matters that cannot be interrupted." Eva began to second-guess her final request.

"No problem, Ms. . . ." the man behind the desk's fingers moved quickly over the keyboard, "Diaz." His eyes continued to scan the computer screen. "Your bags that were dropped off earlier have been placed in your suite. Do enjoy your stay Ma'am."

Hmm. I wonder what Leeds' taste in clothes are like. She shuddered. *Maybe I don't really want to find out.* "Thank you," she said. "Which way to the elevators?" The man behind the desk pointed to her left.

Eva sighed as she turned towards the elevators. *Well, he had to look up my name. Maybe, just maybe, I can be invisible here.* She walked toward the elevator bank and pressed the elevator call button. The lobby was relatively empty and no one else was awaiting an elevator. She glanced over to the front desk. The man she had just spoken to was busy typing something into his computer and Eva thought about slipping around the corner but stopped herself. The elevator arrived and Eva took it to the sixth floor. As she neared her room a young couple emerged from their room down the hall and began walking toward the elevator. They were laughing and Eva slowed down. As they neared, she bent over to adjust her shoe, turning away from them so they would not be able to see her face. They passed and quickly got into the elevator that Eva had just occupied. As the doors closed she stood back up. She found her room and slid the card key into the reader. She heard the audible sound of the lock disengage and she opened the door. She held it there for a few moments then re-shut the door. *There,* she thought. *If they can monitor room use, they'll think I went inside.* She looked up and down the hall. She did not see any cameras, but she knew that meant nothing. *With any luck, Leeds chose this place because it's indiscreet.* She crossed her fingers as she worked her way down the far side of the hall towards the staircase. About halfway there, a door was open on the left side reveling a housekeeping service elevator and store room. Eva noticed a housekeeper's cart outside a room far down the hall from the way she came. She slipped through the door and partially closed it behind her. She pressed the elevator call button, tapping her foot impatiently. A bell sounded indicating the car had arrived and Eva stepped to the side, hoping to avoid being seen if anyone were inside, but the car was empty. She stepped inside and directed the elevator to the ground floor.

The doors opened to a well-lit room. Active laundry machines were lined up across the room before her, their operation adding a droning hum to the scene. To the left she spotted two steel doors with circular windows. *Kitchen,* she thought. She looked to the right and let out a quick sigh. Another steel door. This one solid, with a push bar exit latch across the middle. She took a quick glance to the left and right. *No one.* She quickly ran to the door, and gently pressed the push bar. She slipped through and gently let it close.

She made her out of the alley toward the main thoroughfare and flagged down a cab. "The bus station, please," she directed the driver.

The Number 910 bus took a little over ninety minutes to reach Tel-Aviv from Haifa but Eva felt as it if were only moments. She had tried to fall asleep but her racing mind made that task impossible. In Tel-Aviv, she made her way to the Central Shopping District. She walked among the many hotels in the area and stepped into one of the small boutique hotels.

"Yom Tov. I am wondering if you have any rooms available for tonight?" Eva asked of the man standing behind the desk.

"Yom Tov." The man studied Eva up and down. "An American, it seems. Yes, I do have a room," he said without taking his eyes off her. "Two hundred twenty-five shekels a night," he responded, smiling stiffly.

Eva forced a smile in return. "Of course," she said. She pulled three one hundred shekel notes out of her pocket.

The shoulders of the man behind the counter seemed, to Eva, to have dropped slightly. His smile took on a more natural appearance. "I will need to see an identification, Ms—." He looked at her, questioningly.

"Ms. Di—" She stopped herself abruptly. *Shit, Eva. If you're not going to think this through you're gonna be in real trouble.*

"Ms Dee?" the clerk asked, regaining some of the apprehensiveness Eva had thought he had so recently shed.

"Yes, Ms. Dee. D-E-E. Dee. Joan Dee." Eva slapped both hands from her front pant pockets to the back and back to the front. "Ah, yes. Identification." She tried another smile. "You see, it seems that I don't have any identification on me. I . . . I was in the Central Shopping District when the air raid sirens went off. I remember running, following the crowds towards a shelter, but that's all I remember until I awoke at the University. They checked me out and said I was okay, but my purse did not make it to the University with me. I've tried to retrace my steps and have filed a report with the police, but no luck yet. Now I just need a place to lie down for a little while, take a shower and clean up. It's been a crazy day."

The clerk sighed and nodded in agreement. "I'm sorry, Ms. Dee. I didn't mean to be rude, but there have been some strange things going on, and I'm not talking just about the attack. I, too, remember entering an air raid shelter when suddenly I couldn't breath, there was a raging heat and then everything went black." The clerk looked side to side, as if to determine whether anyone else was in earshot. "I awoke in the University as well. It was not a hospital room. Just a room with a man asking me questions. I do not believe he was a medical doctor. He said that the shelter had suffered a ventilation problem and had been filled with too much carbon dioxide. He said almost everyone in the shelter blacked-out as a result. He asked a few more questions and told me I seemed okay and could leave." He took another quick look side to side and leaned in a little closer to Eva. "When I returned here, everything was exactly as I had remembered right before the sirens went off."

Eva shook her head side-to-side. "I don't understand."

"Everything, exactly as I last recall. Did nothing move from the rush of people? Any papers slip from the desk? A chair become overturned from a guest rushing out?" He looked around. "Perhaps not, the hotel is not very busy this week. But," he paused, "at the University," he paused again, "the soldier who was standing in the hallway when I left, an American soldier," he lowered his voice. "I saw him walk past this hotel a short while ago. It does not make sense." He straightened out and pulled back from Eva.

"I am sorry. I do not mean to scare you." He looked nervous to Eva, as if he had told a secret that was not meant to be spread. "Yes, Ms. Dee. I have a room. You do not need to show identification." He typed a few keystrokes into his computer and handed Eva a key and some change. "Stairs are to the left of the elevators, which are often slow." He turned away and Eva had the distinct feeling he wanted her to leave. That his indiscretion was an error that he hoped would vanish with Eva.

"Thank you," she said and quickly headed toward the stairs.

Eva had made better time than she had expected. She still had time before Clay was supposed to call so she quickly undressed and drew a hot bath. She was relaxing in the tub when her phone rang. The number displayed was that of her own apartment in Chicago. A knot grew in her stomach and the water suddenly and unexplainably lost its ability to warm her tired body. *What if it's not Clay?* She thought about letting it go on ringing, letting the call go to voicemail. She sat up, exposing her upper body to the chill of the room air. *There's no way Leeds could know I'm not in Haifa. And how could he get back to Chicago so fast?* She slid back into the tub hoping to warm the cold empty feeling growing in her stomach. Her mind started to race. *Those fucking Niklases better not be in my place!* The phone went silent.

Eva closed her eyes again and let out a sigh, trying to relax when the phone sounded again. She reached for it and the knot in her stomach tightened.

"Hello?"

"Eva. Hi. It's Clay. Is everything alright?"

She let out a sigh. "Hi. I'm sorry. With everything that's going on I . . . well, I . . . oh, it doesn't matter." The eerie feeling that she was not as safe as she had hoped passed over her again. "It's all a little confusing out here." She thought for a moment. *What can I say? What should I say? What have you heard?* These were the questions she wanted to ask. The thought that Leeds was listening stopped her from doing so. She had a sudden urge to hang up. *Come on, Eva, think of something innocuous to talk about.* "How are my plants?"

"What? Are you okay?" The question had caught Clay off guard. "I've heard some pretty chilling things about what happened out there. I don't know who or what to believe, and I wanted to speak to you, privately." he instinctively looked around. "Your place is still safe, isn't it?" he asked.

"I don't know anything anymore," Eva responded.

"What the hell happened out there?" Clay asked again.

What the hell happened? She considered how she would respond. *What happened is that the U.S. egged the Iranians into a nuclear attack on Israel so we could test mass replication. We risked the lives of millions of innocent people for a fucking experiment. To find out if it's possible to repair all the damage and clean up any nuclear fall-out. To find out if the victims themselves would even know they had been killed. God, I don't know whether to laugh or cry! We just proved that war is utterly and completely senseless . . . that life cannot be erased so easily. It should be shouted out from every corner of the world. Science has finally allowed us to make war and killing ineffective. One nation no longer has ability to impose its might on another. I don't know if we can like one another, but since we cannot hurt one another I guess we'll be forced to live peacefully. What choice is there? War is futile. Kill me this morning and I'll be back this afternoon. You can no longer threaten me . . . harm me.*

But this was done in secret. And it's obvious that Leeds has other ideas about the technology. We have the ability to stop war and killing, but we're keeping it a secret!" Tears began to well up in her eyes thinking of the horror. *And I am helping.*

"There are things going on that could change the world. Change it for the better, but that's not what's going on," Eva finally spoke. "I've got so many things to say, things people should know, but I don't think we should be talking. Not over the phone anyway."

"You've got to get out of there," Clay suggested. "Come back home. You can take it to the press. Spread the word. Once it's out, it'll be too late to stop it."

"No, Clay, you're wrong. No one listens to the news anymore. They listen to what their iMeme tells them. Who do you think puts all that information out there in the cloud? Sure, there are some independents out there, but no one listens to them. And the government has the power to filter the cloud – to stop whatever information they don't want out there." She paused. "No, that would likely just get us killed."

Clay paused. "I guess I hadn't thought of that."

Another pause. Eva's mind began racing again. "Listen," she said, "I should go. I've got a few things to get done over here, but I expect to be back in Chicago next week. Can we meet? I'd love to see you."

"Sure, you just let me know and we'll get together."

"Goodbye Clay."

"Goodbye, Eva. Be careful. Please," he added.

Eva disconnected the call and set the phone down. *It's not safe here anymore. Leeds is done with me, I've got to get back to my place to get my things and get out. While I still can.* She lay in the warm water thinking of her next steps. An image of Clay popped into her mind and she felt her nipples contract and harden with excitement. It suddenly struck her that she had feelings for Clay, feelings which ran deeper than she had known. Of course, she had always loved him in a sense. Their relationship had grown from a silly teacher-student type of acquaintanceship to what Aristotle called a pleasure-based friendship, where each seeks the company of the other because they derive some pleasure from it. But that was not right. She found Clay to be someone she would rely on. She did not have to worry that he would not be there for her in a time of need, now, when things were getting very complicated and, perhaps, very dangerous. Clay had become what Aristotle called the perfect friend. Someone who would not turn their back on you.

It had been a long time since Eva had been with anyone and she felt the sudden need to be loved. To be held in someone's arms in a tender embrace. But Clay was married. She blushed notwithstanding her aloneness, feeling like a silly little school-girl with a crush on her teacher. Nonetheless, she continued to think of him as she lie there, surrounded by the warm gentle softness of the water, her naked body tingling with arousal.

CHAPTER TWENTY-ONE

North Korean troops have amassed along the –

Stop the news, Clay directed his iMe. He looked around Eva's apartment. He wanted a drink. "Vodka on the rocks," he called out. Silence. He paused and realized that no drink would materialize. He envied her lack of a replicator. Life was not so easy for Eva; not so soft, but somehow, perhaps, more real. He thought of the bottle of vodka Eva kept in the freezer. He opened the door to the freezer and found several bottles filled with clear fluid. He pulled out a few ice cubes and poured himself a drink.

What was going on over there? Those men from the HAM radio - one of them said he saw mushroom clouds, yet news reports state that damage was minimal. Iran claims, admits openly, that it launched two nuclear warheads towards Israel, yet the county, seemingly, avoided tragedy. And how was it that only the Al-Aqsa Mosque was in ruins? Why has the U.S. refused to allow the Replicator technology to move across the globe? Wouldn't that have eased tensions across the globe? Why allow these disputes to simmer when the whole thing can be diffused and presumably put to bed by the equality that nanotechnology could infuse to the world community? Clay let his mind start to race. *Would it be so dangerous if the whole world had everything it needed? Of course, war would end. As would, perhaps, government. Yes, government. There was still the base need to control. A primal desire for power. God damn them!* Clay walked into Eva's bedroom and was briefly startled as he looked in the mirror. He still could not get used to being young again. He thought that after two or three months he would grow accustomed to it, but he did not. He hoped that after the passage of enough time it would become second nature, but still often was caught by surprise every now and then at what he had become, or, rather, what he had reverted to. He sat on the edge of her bed and looked at the items on her nightstand. A lamp. A copy of *The Short Stories of Ernest Hemingway* that he had loaned her. *When? Almost a year ago,* he thought. *Perhaps more?* The bookmark showed that Eva was just past the half-way mark. Next to it, a picture of a group of people. Clay picked it up and noticed that the picture was that of Eva, her parents and Lillian and himself. It was taken at Eva's graduation, just prior to the introduction of the Replicator. Eva looked much the same, eyes bright and smiling. Miguel and Jen, the proud parents standing beside their recently graduated daughter. And he and Lillian, looking every bit as proud as her real folks, standing on the far edge. Lillian looked several years older than she did now. Traces of lines on her face, evidence of a small pooch forming, notwithstanding her rigid exercise regime. Lillian was looking into the camera lens, all smiles. He moved the picture a little closer to his face and studied himself in the photo. Evidence of grey at the temples. Clearly not in the same physical shape he was in today, but certainly acceptable for a man of his years, living in a previous age. He seemed to be looking away from the camera, lost in thought, but aware enough of the camera to look in the general direction of the lens. His arm was wrapped around Lillian's waist. Miguel's arms embracing Eva on one side and Jen embracing Eva's waist on the other. Eva was beaming. Young. Proud. Full of potential. She was beautiful. He laid back on the bed and caught a faint, sweet smell as his head sunk into her pillow. He took in a slow, deep breath. Perfume? Shampoo? He finished his drink, lying there staring at the picture.

He tried to clear his mind. *This bed is pretty comfortable,* he noticed. The glass of vodka was resting on his chest. He lifted it briefly. *No way, that's not going to work.* He shifted his weight and sat up. He turned and tossed the pillow up against the headboard and moved to a sitting position. He took another sip of vodka, looked out the window. *What the fuck is going on? Where can I find out?* He let out a sigh. He had nowhere to go and no answers to his questions.

Having failed in his attempt to clear his mind, he let it start racing. A knot crept into his stomach. *The whole world could be going to hell, but who in America would care? Who would know?* The thought worried him and he felt the need to know. Quickly.

Quickly? he chuckled to himself. *In a world with endless free time and endless resources, what is ever done quickly? Back in the day, Matthew would never have finished anything without a deadline. Time was his enemy.* He took another drink from his glass. *Seems it's my enemy now. All this damn time with nothing to fill the endless spaces. Lill seems to have figured it out. In her case anyway. She spends hours at the gym, keeping her firm muscles in shape, in a state that she could neither improve upon nor damage. Not, of course, that I mind that much.* An image of Lillian's naked body flashed through his mind. Her firm muscles, beautiful curves still a turn on. *That I love. But the thought of sitting in a club? Ugh.* He gently swirled his glass, letting the ice cubes spin around the glass. *But isn't that what happens when leisure time is increased? Folks need to differentiate themselves and those with the time figure out ways to do so.*

Funny, he thought. *At one time, when folks labored for a living, really labored - in the fields and forests, just to survive; they were dark skinned from the outdoors and underweight from the difficulties of living another day. The rich, however, having the fortune of others performing the inconveniences of day-to-day life, were able to differentiate themselves - whether by design or perchance. They were certainly pale from residing indoors and fat from an abundance of food. Likely it remained so even into the Industrial Revolution, but somewhere along the way, mankind (or at least the West) reached a point where a large portion of society could be pale and fat. The gym was certainly introduced by the bored and wealthy. How does one differentiate oneself when a large proportion of people enjoy the same privileges? 'Ah ha!' someone must have thought. 'I'll be dark and skinny! The world will once again see my wealth by the fact that my life of leisure is so great that I have time to return to manual labor. But lest one be mistaken, my manual labor is of the kind that is not mandated by my place in the world. It is not for the desire of another, and it results in no benefit to anyone other than myself. This activity, which prevents me from achieving the status of properly satiated, is pure choice. And my utter excess of time - which is so valuable to so many - is such that I can afford to do absolutely nothing at all. So much nothing that I will get up in the morning and go right back to sleep - but rather than in the comforts of my bed - in the outdoors, under the sun so that none would mistake my leisure as a necessity of inclement weather, but a choice available to those who can afford to follow the sun!'*

But now? Shit! The gyms are merely filled with hundreds of people, exercising bodies that need no exercise, bodies that will remain firm even in the absence of activity.

But I'm being too harsh. Lill loves to work out. Hell, I still play hockey two times a week. But I can't bring myself around to the continuous play that many of my fellow skaters undertake. Or sitting all day in one of the many bars and restaurants that continue to operate; mostly by chefs who relish the exercise of their culinary creativity by developing succulent dishes for the replicator to fashion from nothing, and occasionally, I suppose, by a bartender who simply enjoys the company of his guests. That's where the center of society is now. Eating and drinking to excess. With no fear of weight gain and no fear of over-intoxication. Ancient Rome, Clay thought. The nanobots exercising their duties inside one's body were programmed to allow one to get pleasantly buzzed, but would prevent anyone from becoming too drunk or otherwise overly-intoxicated. Clay himself found it pleasant to sit and eat and drink all evening without worrying about whether the previous drink should be his last. And on occasion, Clay would spend his day seeking out an antique store, combing them for some good reading material, some classic that was not yet destroyed or discovered by one of the few people also still seeking books. But the number of antique stores still operating continued to dwindle and those that held an inventory of books was significantly less.

His eyes wondered about the room and landed on the window. The skies were pure blue. *Another beautiful day in paradise,* he thought. He let out a quiet chuckle. *Of course it was another great day. It was always great.* The weatherbots kept the skies clear all day and the temperatures in the mid-70s. Rain was scheduled for off-peak hours, usually between 4 a.m. and 5 a.m. Whether these nightly storms produced lighting was seemingly random. Of course, lightning could always be viewed at local "strike zones", where the bolts of electricity would reach down from the sky in brilliant flashes of white-hot light and ozone. Scientists still had not figured out the consequences of a lack of lightning - what an electrical imbalance between the earth and atmosphere would result in - and in order to maintain the presumed necessary balance, created hundreds of "strike zones" all around the country to ensure that the electrical balance remained in check. He often enjoyed driving out to one of the many strike zones and watching the spectacular shows. Although the randomness of lightning strikes in the nightly storms provided a much more natural and exciting experience, he found himself oftentimes setting his alarm to get up and watch the strike zone events in the middle of the night. He felt the nighttime strikes were far more satisfying than the daylight strikes.

Clay noticed his glass was empty and set the picture back on the nightstand. He watered Eva's plants and noticed that the Mexican Anise was erupting blooms. The cactus had an aromatic bloom that always reminded Clay of evenings in Akumal, when he and Lillian were first married. A couple of times they rented a small cottage on the beach and spent their days snorkeling or scuba diving and their evenings walking arm-in-arm among the several shops, sharing an ice cream or perhaps a churro. The beaches would always offer plenty of free space and the quaint village was home to many small villas which provided a beautiful backdrop for their late evening strolls. He had always suggested that they should purchase a small home there so that they would have a peaceful place to retire and grow old together. Clay felt his mood change to melancholy. *Grow old together*. The phrase seemed to hold about as much meaning as his marriage. He still loved Lillian, but he did not feel the same spark that he had felt before the replicator changed their lives. He recalled that the passion was there even to the time he had been innoculated, but since achieving immortality, things were changed. He and Lillian still spent most evenings together, sharing their time with friends or family, but lately it seemed that they saw friends more and family less. He couldn't remember the last time he had had dinner with his kids. *Hell, I can't remember the last time I've even seen them. What did they do with themselves?* he wondered. No one went to college anymore – knowledge was simply uploaded to one's cerebral cortex. And since no one other than government workers and, seemingly, a select few scientists, of which Eva was one, worked, the rest of the country was filled with artists practicing their crafts and the ever-present consumer, complementing the artists by vying for their wares. The news spoke infrequently about the military personnel, the thousands of troops stationed around the globe, but Clay did not actually know anyone who was, or had children involved, in the military. He could not help but wonder if our armies were now staffed solely with Niklases, acting in unison to carry out some unknown U.S. objective. He thought back. *Was it yesterday I saw Matthew hooked into his V-box or was it last week? How old is Matthew now?* He thought hard. *When was it that Katie was supposed to finish college? Did she enjoy it? Did she finish before they began uploading knowledge? Did she even start before then?* The Replicator interfered with everything. He suddenly felt as if he had awoken after a heavy drunk, vaguely remembering the characters and places of the past evening, but unsure of when he lost track of the night, of what was said or done. *How long ago was I immunized? Since I went to my office? Since I added any value to the world?* A frown crossed his face. *How can I know? There are no seasons any longer.* Living in continual summer-like conditions, he realized, interfered with the ability to track time, the movement of earth on its journey around the sun. *What date was it?* He did not know. *When was the last time I thought about it? Would the iMeme know?* The sudden loss of humanness scared him. *Where am I?* he thought. *What has become of everything?* He leapt to his feet and set his glass down in the sink and departed, almost forgetting to lock the door when he left.

He returned home to an empty house. He grabbed his iMe.

"What is today's date?"

April 23rd, it responded. "What year is it?" Silence. "What is the year?"

A year is a measurement of time approximating the period in which it takes the earth to make one complete orbit of the sun.

"No, what is today's year?"

Today is not a year. Today is the period of time starting from the most recent 12:00 a.m. occurrence and continuing through the next 11:59 p.m. occurrence.

"No, no, no, god damnit! What is the date today, the year and the day?"

Your question does not make sense.

Clay threw his iMeme against the wall and it smashed into several pieces. Just as quickly, the individual parts gathered together and repaired themselves. When Lillian returned home, she found him in the kitchen, throwing rocks through the window, watching the glass shatter and reform. A small pile of stones lie before him on the table, a larger pile lay scattered outside the window.

"That's a little juvenile, don't you think?"

"Oh, huh? Hello. I didn't hear you come in."

"I wouldn't think you would over the continuous sound of breaking glass." He stopped.

"What year is it?"

"What?"

Clay paused a moment. "I asked, 'What year is it?'"

"Why, it's . . . it's . . . . I don't know." A shadow crossed her face briefly, but quickly passed. "But what does it matter anyway?"

"Seriously? You don't care what year it is?"

"No, honey. Why should I?"

"Why should you?" he answered. "Because . . . because . . . well, aren't you curious as to what the date is, or perhaps why we've all seemed to stop caring? I can't get my iMeme to tell me and it seems odd that nothing seems to carry the year anymore. I realized today that I have no idea how old I am, or you, or the kids."

Lillian chuckled. "Is that what's bothering you? That you can't wish me a happy 350th birthday sometime down the line? Do you think folks want to keep track of their age when it is meaningless? Honey, there's no reason we can't live forever. When do you just stop keeping track and concern yourself with living? With the here and now, not what's been or what may be?"

"But why can't I find out the information? Just because you don't want to know doesn't mean that I don't."

"Jesus, Clay. Really?" Lillian's voice lost a little of its softness. "You want to waste your time worrying about what year it is? You've got to start doing a little more. You can't just sit around trying to hold onto the past. Let it go." She turned to go but stopped herself. She stepped behind where he was sitting and placed her hands on his shoulders. She massaged them and Clay felt her strong hands begin to loosen his muscles. He let his head fall forward onto his chest. He felt her warm breath on the back of his neck and then her tongue lick down his neck toward his shoulder. Soft, gentle kisses. She worked her way to his earlobe and sucked gently upon it. He tilted his head so that she could work her way around his neck toward his chin. She spun the chair he was sitting in around and straddled him. Their lips locked in a long, passionate kiss. He stood up, she wrapping her legs around his waist and he carried her to the counter, embracing and continuing to kiss. He let the counter take her weight. He felt her tongue tangle with his and he pulled her shirt off, the two of them stopping momentarily as her shirt came over her head. They quickly resumed kissing, her naked breasts pressed against his shirt. She reached between them and tore his shirt open, revealing his firm stomach and abs.

Lillian slid off the counter and walked seductively out of the kitchen and to their bedroom. She adjusted the shades down and the room darkened. They continued kissing, touching, their hands caressing here and there. Lillian wrapped her hands around Clay's head and then it seemed that she was also wrapping her arms around his chest, rubbing his body, continuing the dance that wrapped their bodies together, becoming as one.

When they were done, Clay was lying on his back, eyes closed and relaxed. He rolled over and looked at Lillian lying next to him. He suddenly sat upright in bed. "What the . . . ?" he began. Lillian was her old self. Hair slightly grey, her weight back to what it was before the PreventAll, her small belly revealing to the world that she had carried several children in her youth.

"What the . . . you're . . . you're your old self again!" He winced. He did not want to use the term "old". "I mean . . . "

Lillian gave him a puzzled look and then reached up and pulled off the virtual reality device she had slipped on him earlier. She pulled him in close.

"What you mean to say is that you're a pathetic moron. For a man who likes to think of himself as sophisticated, you should know that the latest PreVentall update allows people to see others as who they want them to be." She blinked as if losing her train of thought and whispered, "Goodbye Brad." She smiled coyly at Clay and returned to her normal voice. "Admittedly, I'm not sure that it has a lot of use outside of the bedroom, or in front of the the mirror. But it certainly spices things up. If you had the latest update, you wouldn't need the dumb device. Didn't you notice that I had put them on you?" She did not wait for an answer. "Of course not. The sad part is that while I just made love to Brad, you're fantasizing about some old lady."

"Sorry," he said sheepishly as Lillian rolled out of bed and stepped into the bathroom. "Wait. Brad who?" he called after her. She did not respond.

He heard the shower turn on and he leaned up against the headboard. The encounter was a mixture of pleasure and ill-ease, of intimacy and distance, not unlike his experience with the Virtual Fantasy that he and Lillian had tried seemingly so long ago. *I made love to Lillian, there's no doubt about that.* He paused his thoughts. *Who did she make love to? Of course, I was there, it was me and her, my body there, my muscles moving, my body rocking back and forth, but then again.* He slid out of bed and worked his way to the kitchen for a drink. *Who was Brad? And why does it feel as if that really didn't happen between us.* He began to question his perception of what had just taken place. He laughed silently to himself. *Well, there is certainly no way in hell I expected to satisfy Lillian for the rest of her life, whether in the past when age and body would fail me, or in the present, when our marriage could last hundreds of years and boredom and routine would certainly be my betrayer.* He took a long swallow of his vodka. *Why not just be happy? I have Lillian back for now, my old Lillian. And she can have whatever flavor of the day she wanted - was it really so different than her imagining it?* He smiled, but as quickly the smile fled from his face. *When would such fantasy stop being fantasy and cross into infidelity? Would she ever make love to him again? Did she even see him as himself outside the bedroom? The man she married, a man that perhaps she is now unwilling to love?* He pushed the thought aside. He could not think that hard right now. He did not want to.

* * *

When Eva returned to her apartment in Tel Aviv, Lt Cobbs was standing outside her door. Her look of surprise must have amused him, because he immediately began to smile.

"Well, hello Professor. Back from a wild night on the town?" He reached out to touch her face. She took a step back and looked down at herself. She was wearing the same clothes she had put on over 36 hours earlier, much worse for the wear. *And likely me, too,* she thought. It wasn't until she washed her hair and got out of the tub back at the hotel that she realized that she did not have any toiletries. She had to run her fingers through her hair in an attempt at brushing it, but it still looked disheveled.

"I'm sorry, Lt Cobb. I'd love to invite you in, but it's been a long day and I just need to sit down and rest a bit." She placed her fob at the door and attempted to enter the apartment alone. Lt. Cobb quickly raised his arm to the area of the door above her head and pushed the door open wider.

"Invite me in? Why I'd love to. Thank you, Professor," he said, mockingly, as he pushed Eva in and closed the door behind them. Eva caught a glimmer of something in his eye, but she could not yet tell whether it was anger or lust.

"What do you want, Cobb?"

"What do I want? You have me locked up for two days for some bullshit testing and you want to know what I want? You know goddamn well what I want." He stepped closer to her and grabbed her arm. "And I'm going to get it."

She attempted to pull her arm away, but his grip was too strong for her. She slapped his face, but he grabbed her free hand and wrapped both of his arms around her, forcing her hands behind her back. He was pressed up close to her body and she could feel the anger and passion building within him. "Oh, so the Professor likes it rough, huh Eva? Is that how you want it?" He forced his mouth on hers, attempting to slide his tongue in her mouth. Eva bit it hard. She tasted blood and spit the end of his severed tongue out of her mouth.

"Ow! You bith!" he said, more out of surprise than pain. He spit onto the floor and put his hand in his mouth. He pulled it out and looked at his fingers. "I 'hink you bi' par' of my 'hung off!"

He forced her backwards into her bedroom until she fell onto her bed. He was now on top of her. She struggled to get from beneath him but it was no use. He slapped her across the face. The stinging pain made her stop struggling for a moment. He put his fingers back to his tongue but by that time, the PreventAll bots had completed all their repairs. "Oh, you do like it rough, don't you?" He smiled as he unbuttoned his shirt with one hand. "Two can play that game." He finished with his shirt and tore hers open in the front, exposing her breasts. He pinched at her nipples roughly. She started to struggle again. He repeatedly slapped her across the chest. Red welts appeared where his fingers struck. He bent over and began to suck on her breasts, gently at first, then biting down hard. Eva stiffened in pain and then quickly tried to move her arms. Cobbs' knees were placed firmly on her forearms, trapping them beneath her. She tried to roll over. "I think we've got a fighter here," he stated as if providing a play-by-play to an invisible audience. He reached behind his back and Eva saw that he had pulled out a zip tie and placed it between his teeth. She struggled harder as he pulled one hand out from behind her and forced her to roll over. He quickly pulled her arm up high behind her back and placed the zip tie around her wrist. She continued to struggle and he pulled her arm further. She let out a scream. The pain was excruciating and Eva thought that he might pull her arm out of its socket. Cobbs only smiled. "Nice touch, you act like I'm really hurting you." She stopped struggling and Cobbs grabbed her other arm and zip tied it to the first. "What's the matter, Professor? You're not ready to give up already, are you?" Eva lay there, contemplating her next move. She was helpless.

"You bastard!" she hissed.

He slapped her hard on the rear. "That's right, Professor. Get riled up!" He reached his hand around her waist and unbuttoned her pants and pulled them down, exposing her buttocks.

"Ouch!" Eva screamed as Cobbs bit her ass. She kicked her legs up and back.

"That's right, baby!" He slapped her bare behind again. He started laughing. "You've been a very bad girl, Professor. Very bad!" He slapped her several more times before he flipped her over. Eva's pants were around her lower hips and Eva continued to kick. He sat on her legs and reached down to strip her pants all the way off when he suddenly stopped. Eva had blood streaked across her chest. Her left nipple was bleeding. "What the fuck?!" Cobbs got up. "You're fucking bleeding!" Eva looked down and saw a fresh swell of blood rising from her breast. "Your still fucking bleeding," Cobbs repeated. He looked at her, dumbfounded. "You god damned freak!" He grabbed his shirt and tried to cover his bare chest, as if he were trying to protect himself.

"I . . . I . . . ." she started.

"You're a fucking BAMF! You lying, sneaky bitch! Wait until the Colonel hears about this. And when he does, you're not going to be so fucking high and mighty anymore, Professor. God damned BAMF!"

"Cobbs!" Eva started.

But Cobbs wasn't listening anymore. He continued to mumble. His eyes wide, a look of fear and disbelief on his face. He finished grabbing his things and scanned the room as if expecting someone or something to jump out from hiding, then turned and raced out of the apartment. Eva heard the door close behind him. She lay there for a moment. Her body was bruised and sore, but she managed to slip her hips through her bound arms and then her legs, so that her hands were now in front of her. She went into the kitchen and retrieved a knife from a drawer and held it between her hands and slowly moved it back and forth. After about five minutes she was able to saw through the zip tie, freeing herself at last. Her whole body ached and she slumped to the floor and began to cry.

When she awoke it was dark outside. She did not remember climbing back into her bed. She stood up before the mirror to inspect herself. The blood on her left breast was dried. She had bruises across her chest where Cobbs had abused her. She turned to the side to inspect her buttocks. They were red and tender. She could see a bruise in the shape of teeth marks forming on the right side. She cleaned her cuts and pulled out a vial of ReGenbots to apply to her skin. The vial was empty. She looked around for another vial but failed to locate one. "Oh well," she said to herself, knowing that while the ReGenbots would mask the surface wounds, they did little to heal the deep bruises. She applied some cosmetics gingerly to her sore body, attempting to conceal the wounds as best she could. She would simply have to deal with the pain - a task she had learned to to many times in the pasts. She grabbed a duffle bag and started to stuff some items into it, items she felt she may need.

She picked up the phone and dialed. It only rang once before being picked up.

"Good evening, Professor Diaz."

"Hello, Colonel Leeds."

"What can I do for you?" His tone was steady and guarded.

"I'm going home. Tonight," she added quickly.

"Can I ask why?"

Eva felt like Colonel Leeds was toying with her. "I think my work here is completed, sir. I miss my family."

"And is that all?" he queried.

She sensed something was not quite right. "Yes sir," she said, trying to sound truthful.

"Are you sure?"

"Quite." She tried to figure out where he was going with the conversation.

'Well it seems I have a certain Lt. Cobbs here who suggests you may have different reasons."

Eva's mind began to race. *It was DARPA who would not permit me to be inoculated. I'm not even sure I would want to be, but the decision was out of my hands.* No iMeme. No Replicator. No PreventAll. Her position was too important. The government feared that the scientists working on the top secret projects could too easily be exploited and forced to reveal secrets, or perhaps desire wealth and power of their own and sell state secrets to foreign entities. She thought about the warning she was given against disclosing her BAMF status. *It was the government's decision. I'm not allowed to let anyone know. I'm not really going to be punished now because I am a BAMF, am I?* Her stomach knotted up. "I . . . I . . . "

Eva heard Lt. Cobbs in the background. His voice was easily heard over the phone. He was screaming, "Tell him, Professor. You were bleeding! You're a god-damned BAMF! Go ahead, admit it!"

"Sit down, Lieutenant! I will not have you acting in such a manner!" Col. Leeds spoke firmly but without raising his voice.

"I apologize, Sir! My action was unacceptable," Cobbs said, his voice lowering, becoming harder for Eva to hear. "You understand, though, I'm sure, Sir." Eva could sense the submission in Cobbs' voice, but at the same time a certain cockiness in his final sentence

"Don't be so certain, soldier." A pause. "Professor," he began speaking again before Cobbs managed to question the colonel's last statement, "would you like to rephrase your answer?"

Eva's mind began to race. *What did Cobbs say? Certainly he did not walk into Col. Leeds' office and tell him he tried to rape me. He couldn't have.*

"I'm sorry, Colonel. I'm not sure what you're asking here," she stated. Eva began to sweat. *Did he know? What information was Colonel Leeds privy to? What if he did not know? Certainly he would contact his superiors before he took any serious action. Did he have superiors?* She continued, tentatively, "Yes, sir, it's true. I was bleeding. Certainly you were aware that I was not inoculated . . . ,"

"Certainly I *am* aware. Perhaps you would like to elaborate on how the Lieutenant came to acquire this information? It seems I am unable to calm Lt. Cobbs down long enough to get an answer out of him." There was a moment of silence. Either, Cobbs wasn't willing to start explaining how he came to his recent revelation, or he was prevented from doing so.

"Ahhh, of course, Sir. I'm sorry. It's late." She paused to buy a little time. "Lt. Cobbs came to my home earlier today." Her mind started to race. "It seems that he wanted to let me know personally that his Lab 3s and Series VI protocols came back normal. Sir," she added, continuing to try to formulate a believable response. "I had sent him for testing a few days ago after he indicated that he wasn't feeling quite right." Eva was tempted to tell the Colonel what took place earlier today, but she was not sure what was happening, what game the Colonel was playing. A certain uneasiness rushed over her. *This isn't quite right. Something is going on here. Something serious. This isn't just about Cobbs attempting to rape me, if that is even in play here.* She held a revulsion for Cobbs, but her compassion rose up. *I do hope that son of a bitch get his due, but . . . but –* It seemed to her that this was not merely about punishment, but something much more. *Maybe even fatal,* she thought.

"I invited him in for some tea and when I was, uhh, cutting lemons, I, um, accidentally sliced my index finger." A story was rapidly forming in her mind. "It was a pretty deep cut. I put some ReGenbots on the cut to repair it, but I didn't realize at the time that I had also cut my ring finger, too. I served the tea and Cobbs and we talked about his medical testing. He stayed for just one cup and then left. It wasn't until later, while I was cleaning up, that I noticed the cut on my ring finger. It probably had been bleeding the whole time we were talking. I guess it was during our conversation that Lt. Cobbs noticed my finger continued to bleed." Her pleasure with her story suddenly turned. *God, I hope Cobbs kept his fucking mouth shut.* She felt the need to explain more, cover her ass if he had already stepped too deep into the shit that seemed to be surrounding her.

"He's very . . . observant," she added. "He didn't say anything before he left, sir." She held her breath, unable to think of anything else to say.

"I see. Thank you, then, Professor." Leeds voice remained calm. "I expect you'll be on the next flight back to the States. Tickets will be waiting for you at Gate Aleph-3. Your identification cards will get you through security checkpoints – I have already notified the IDF."

"But . . . but . . . . You're not just going to let her go, are you Colonel, Sir?!" The hysteria in Cobbs' voice was easily heard over the phone. "She's a BAMF, Sir! She a danger to society, she's—" Cobbs' tirade ended abruptly. Eva had heard the sound of a scuffle and the muffled protests of Cobbs.

"One more thing, Eva," the Colonel said, his voice still calm but with hint of forced gentleness. "The events of today, regardless of how they occurred," he paused. "They never happened."

Eva remained silent. *Did Cobbs tell him the truth of what happened?* A terrifying thought crossed Eva's mind. *Had Leeds simply downloaded Cobbs' memories? It would be a simple task. Fuck! Why is it so difficult to remember what we do? He played me.*

Leeds continued, his voice remaining calm, as if nothing odd had occurred. "I would hate to have to repeat the conversation I am about to have with Lt. Cobbs with anyone else." The line went dead.

Eva hurriedly stuffed the last of the items she thought she may need into her backpack. She was rushing out the door when she suddenly turned back into her apartment. She quickly went to her bedroom and opened the nightstand drawer. Beneath several magazines lie *The Miraculous Journey of Edward Tulane.* It was the most recent book that Clay had loaned her. She stuffed it deep into her backpack and ran out.

* * *

"I would never deny a man his guilty pleasures, Lieutenant," Leeds said placing his hand gently upon Cobbs' shoulder Cobbs was strapped down on a gurney, flat on his back. His mouth was taped shut. A look of fear dwelt in his wide eyes. "But Professor Diaz is off-limits. You clearly should understand the importance of her in my work, in our work, Lieutenant. We can ill afford to lose her." Leeds was walking slowly around the gurney, not looking at Cobbs. He stopped at the head and leaned closer to Cobbs' ears.

"Tell me, soldier. Do you know what we do when someone crosses a line?" Although Cobbs' head was strapped down over the forehead, he was able to shake it back and forth slightly. His eyes remained wide and fearful. "Of course not," Leeds continued, once again, pacing around the gurney. "Let me explain. A good soldier enters battle knowing his enemy, what he must face. We can't have you unaware of what is going to happen. I wouldn't want you to be afraid of the unknown." Leeds paused at the foot of the bed, and finally looked into Cobbs' eyes. A smile spread across his mouth. "No, I want you to be afraid of the known, of what you are going to face."

Cobbs began to wriggle and a muffled scream rose from his sealed lips.

"Your efforts are futile. You will not free yourself." Leeds began his circular pacing again. "You may want to save your energy, Lieutenant Cobbs. You see, we are about to reprogram your PreVentall bots. Shortly, they will no longer serve to simply heal your cells. Rather, they will undertake quite a similar and opposite effect, and destroy your cells while keeping them functioning. But don't worry, Lieutenant, this will not take place all at once. That would be much too quick and easy. No, I think we will begin with your integumentary system. Have you ever been burned, Lieutenant? I mean, really burned?" Leeds did not seem to care for an answer. "They say it is quite painful when the skin is burned off and the nerves are exposed. Yes, that is where we'll begin. Skin, hair, nails. Not all at once, mind you. Just the outer layers of skin to begin with." Leeds stopped walking briefly. "I wonder what it would feel like to have road rash over the entire body all at once. Quite unpleasant I would imagine." He began his pacing again, Cobbs, unable to turn off his ears, was stricken with a fear that froze him in place.

"Soon enough - or, perhaps in your case, Lieutenant, not soon enough, but let us ignore the dirty details, shall we? Soon enough, we will remove your skeletal system. Imagine what you'll look like, all pink and soft, wriggling in pain, your organs sloshing around in your skin, spreading across the table. One must wonder if there would be any relief if we removed all your skin, if you were simply allowed to ooze out, relieving some pressure on the tender skin. In time, that will happen, but we do not want to rush matters. The muscles and ligaments will be next. Yes, you may be thinking, that without your muscles to circulate your blood, you will soon pass. But certainly the thought that the bots in your remaining cells are capable of extracting oxygen and nutrients from the environment and supporting continued cellular life has crossed your mind? The real fun starts when you are nothing but the nervous system. Mostly nerves and thoughts. I will not be here that many days, so I cannot say what the Niklas' may do. They have told me that a cat will apparently believe that what is left of you is nothing more than a ball of yard, something to claw and scratch, and perhaps nibble upon. Perhaps they have some other plans for you, I cannot say. I will try to stop by sometime next month for a chat. Oh, excuse me. I do apologize, but it is too messy to keep your mouth and vocal cords functional during the process. And as you can imagine, your screams would simply become too much after a week or two to allow us to finish our work with you, so it is I who shall do the talking. I am certain you understand. But do not fear, we have arranged to allow you the use of your eyes and ears, so you will not be completely unaware of your environment. I will be sure they secure your eyeballs so that they are not continuously rolling around while we visit next, perhaps in a month or two. I do so look forward to our times together."

A panicked scream rose from Cobbs throat, as his skin began to redden. A thin film of blood and plasma began to coat his entire body.

"Oh, I do so wish I could stick around, it looks as though things may start becoming interesting. But alas, duty calls. Do take care." Leeds placed his hand on the doorknob and turned once again to Cobbs before opening it. "Oh, and if Professor Diaz happens to stop by, please remember, you *will* show her the utmost respect. She is mine." With that, Leeds exited the room.

CHAPTER TWENTY-TWO

The all new BeMe *app is here! Amuse your friends, make your neighbors jealous, relive those great memories. It's all possible with the all new* BeMe *app! You can now transmit your own thoughts, memories and dreams to others. Imagine the possibilities! Your friends will have to believe you now - because they will live it through you. Included in the latest iOs PreVentall Release 3.1. Try it today!*

When Clay returned home, Lillian and the kids were all in the living room, laughing.

"What's so funny?"

"Oh my god, dad! You'll never believe it! Matthew pulled the funniest stunt ever! Show him, Cal." Katie turned to her brother.

Matthew directed his iMeme to share his experience with his father. "Dad, you'd better sit down."

Clay complied, not sure what to expect. "Send it to Dad," he commanded his iMeme. Matthew's face formed a scowl. "Why can't I send this to Dad?" He commanded his iMeme to try again. "What? A 404 error?" He rolled his eyes. "Dad, are you seriously telling me you haven't upgraded your PreVentall? You're still running 1.0? What's the deal?"

Clay smiled sheepishly. He shrugged his shoulders. "Yeah, I guess so."

Katie threw Matthew a look that said: *Don't harass him too much.* "Dad, when are you going to pay attention and keep up with the world? It's been announced for the past few days – they've figured out how to share experiences, real experiences." Katie started to speak faster, as she tended to do when she was excited. "You can capture any point of your life and upload it to the cloud. Then others can share the experience but rather than reading about it or seeing it, you actually live it - it's like you're in someone else's head. It's crazy!"

"You know your father, kids. He's only going forward kicking and screaming," Lillian said, directing her comment to Matthew. She smiled at Clay.

Clay was reminded of the Christopher Walken movie *Brainstorm.* It had some pretty disturbing scenes of shared experiences and he worried that people did not really understand the ramifications of the technology. "I have a hard enough time living my own life, so it's all good with me," Clay said to his family, winking at Lillian.

******

The motion of the aircraft startled Eva into wakefulness. She opened her eyes, briefly unaware of where she was. The darkened cabin was filled with faces of several different ethnicities, some lit by the light of video monitors in the seat back in front of them, others shrouded in darkness, attempting, like Eva, to secure a few hours of sleep. She felt a disconnection from being the only American on the airline. The flight, like all flights to America, was mostly empty. Persons with the means and connections were able to travel to the States, mostly for government matters, but some for pleasure. Their visit would be closely monitored and freedom of association was limited by the need to protect PreVentall and other nano-technology from exiting the country. The plane shook violently as it entered more turbulence and a few passengers let out a gasp of fright. *Ding!* The 'Fasten Seatbelt' signs flashed to life.

"I apologize folks. We seem to be experiencing some turbulence here. We will attempt to increase our altitude to get above this rough patch, but ask that everyone return to their seats until we can smooth out our flight." The captain's voice had the air of mock cheerfulness, a clear attempt to reassure the passengers. It was not present when he added, more softly and in a serious tone, "Flight crew please return to your seats."

The skies outside the window were dark and Eva glimpsed flashes of light indicating lightning. She pulled the small blanket higher around her neck. Her body ached all over. Her cheek felt swollen and throbbed slightly. She hoped no bruises were visible, but had little time to ponder the issue before she struggled to fall back into a troubled sleep.

* * *

Clay knocked on the door again, but there was no response. He placed his ear on the door to see if he could hear any sign of activity inside. He listened closely but could hear nothing. His hand slipped into his pocket and he fingered the key to Eva's apartment. She had called earlier and asked him to meet her. He gripped the key between his fingers as an uneasy feeling rose in his stomach. He had decided to let himself in when he spotted a woman walking down the hall, her gait a little labored. She had a scarf wrapped around her head and a large pair of sunglasses resting on the bridge of her nose. She stopped in front of Clay and without looking at him, opened the door. She slipped inside, leaving the door open.

"What are you doing? Get inside!" It was Eva's voice. She pulled him into her apartment.

The sunglasses were sitting on the small table in the entrance way and she was removing her scarf when he turned to her again. Clay almost did not recognize her. Eva had sounded upset on the phone but he certainly didn't expect to see her like this. He never expected to see anyone like this, at least not anymore.

"What the –?" he started.

"Oh shit!" Eva gasped. She let the scarf fall to the floor as she moved quickly to the bathroom and flipped on the light. "Jesus."

Clay snapped to and followed her to the bathroom. "What the hell's going on?" He was standing behind her, both of them looking at Eva's reflection in the mirror. Clay noticed that the white blouse she was wearing had the top two buttons undone, showing some cleavage and evidence of her recent encounter with Lt. Cobbs. The skin of the left breast had a dull bluish hue. *'Odd,'* Clay thought as he moved his gaze further down Eva's body. She was not wearing a bra and her nipples protruded from beneath her blouse, which was only partially tucked into a pair of black jeans, low cut and highlighting the curve of her waist. His gaze returned to Eva's face again and he noticed that she had bags under her eyes. A wisp of gray hair hung noticeably on her left side.

"Well that may explain it," she mumbled to herself, flipping the gray hairs back. Clay was still staring at her, intently.

"I'm sorry you had to see me like this" she said, turning around to Clay. "It's been one hell of a week," she said softly, more to herself than to Clay. She turned around pulled open one of the drawers and pulled out a vial of ReGenbots, but Clay grabbed her wrist before she cold open the container. He turned her around, staring at her, intently. He moved his gaze from her eyes to her hair and back again. "It's a long story, Clay. I'm not really inoculated. I'm . . . I'm . . . -"

She leaned forward and kissed him. Instinctively, Clay withdrew, but she pulled him back near as she began to kiss him again. His tongue moved past her lips, exploring her mouth as they held each other more tightly. She backed up slightly, continuing to kiss him, but moving her hands along the front of his shoulders and to his chest. She struggled to unbutton his shirt, but failing, simply grabbed it and tore it open and then pulled her own blouse open. She tried to pull him closer, but then leaned into him with her mouth, pulling her body back slightly, keeping the pressure between them light, so as not to irritate her tender body. Her breasts felt warm against his chest as they held each other a little closer. They continued to kiss as Clay's eyes closed and his mind began swirling. He moved his lips along her cheek and down her neck. She tilted her head to the side and let out a small moan. He felt her fingers slip under the waist of his pants as she started to unbutton them. He let his hands slide down her waist to her hips and was surprised when he realized that she had already lowered her pants. His hands continued down and around and he gently followed the line of her underwear. They were making their way towards Eva's bedroom, shuffling along as the result of being hobbled by their not-fully removed pants. Walking backwards, Clay stumbled and nearly tripped. They parted lips and he finished removing his pants as she stepped out of her garments and stared at him. She gave a smile and immediately wrapped her arms around Clay's neck as she jumped up. Clay held her as she wrapped her legs around his waist, easing herself around him. He carried her into the bedroom as she continued to kiss him. He laid her on the bed and moved his lips down her neck and to her breasts. They felt somehow familiar as he gently licked around her damaged nipple, before moving down her belly and further. He worked his way back to her mouth and they wrapped their arms around each other. Neither said a word until they were done. Eva was still breathing heavily. "I'm . . . I'm sorry. I didn't mean to . . . ." She pulled the bedspread around her and moved off the bed toward her closet to grab a robe that was hanging inside. As she put it on she kept her back to him.

Clay sat up and stared at her back. He grabbed his underpants and sat on the bed putting them back on. "No, it was my fault. I should not have let it get this far. I . . . I'm just not sure what came over me. When I saw you there in the mirror, with some grey hair, looking . . . ."

"Like a real person?"

"No. Like . . . like," but Clay couldn't quite place his finger on the shadow of thought that eluded his grasp. "Yes," he stated, giving in to the fact that he was not able to capture his thought. "Like a real person. Somehow I . . . I . . . I don't know. I can't explain." He stood up and Eva turned around. She had tears in her eyes.

"Jesus, Clay. I'm so sorry. I never meant to do that to you." She looked to Clay and then quickly away, "to Lillian." A tear ran down her cheek.

"Lill," he said softly, "what-," he broke off. *What did I do Lillian? With Eva . . . with us?,* Clay thought. He started to feel confused. *Do I still love Lillian?* He thought a little more. *Of course I love Lillian. The mother of my children and my wife who . . . who . . . .* His mind stopped for a moment as it changed direction. *Who is she? She is Lillian, of course, but she isn't quite Lillian anymore, is she? She's not the same woman I fell in love with and married.* His thoughts wondered into pure emotion, *I so love that woman I married,* but quickly returned the questions that were infecting his brain. *But who is she now? Who am I? Where are we?* He looked closely at the back of his hand, turning it slowly and examining the front. Eva had sat down on the bed, her back to Clay. He studied the curve of her spine. *What can I say to Lill? What have I ever said lately? When was the last time Lill and I spoke, really spoke? When was the last time I spoke, meaningfully, to anyone?*

Clay watched as Eva stood up and went towards the bathroom. *My God - what just happened?* he thought. *Did it really happen?* His head began to spin and he wasn't sure what was real anymore. He looked around, trying to recognize his surroundings. He had a sudden feeling of deja vu. *Last time Lillian and I made love, where were we? Who were we?* He tried to remember. *Was that even real?* He looked around again. The sound of running water came from the bathroom. "Lil?" he said softly. He shook his head. *No, that's Eva in there. I think.* He looked around, still unsure, half-expecting to see the Virtual Fantasy goggles lying beside him. His thoughts returned to Lill and the last time they had fooled around. *I think we had intercourse, but was it us?* The goggles had seemingly become routine in their love-making. *Was Lill making love to me, or to "Brad"?* He could never know. He paused in his thoughts. *Was I making love to Lill?* He shook his head. *Who have I been making love to?* He contemplated the question a little harder. *No, it hasn't been Lill. It hasn't been her at all for some time. It's been a woman who was . . .* His mind struggled to complete the thought. *A woman who was . . . real?* He shook his head again. *What is real? Natural faults? Wrinkles, bags under her eyes, grey hair, maybe a pooch? Eternal youth? That's the new real, isn't it?*

It struck him that he no longer recognized himself, his relationships in the world. *Where am I? Where did we, Lill and I, go?* They no longer shared their lives. Each running about doing their own thing, occasionally sharing a meal together, sometimes enjoying one another's company. But no longer sharing a life. *Hell,* his thoughts returned to his earlier wonderings, *we don't even make love to each other. We simply use each others' body to underlie the fantasies that play out in our eyes and minds.*

Clay stood up and was pulling his pants on when Eva returned. Clay turned to look at her. Her reddened eyes and flush cheeks made it clear that she had been crying, but that was not what struck him. His gaze intensified. Although puffy, her eyes seemed younger, firmer. The bags that had been their earlier were gone. His eyes moved down the vee of her robe. He did not see the earlier tell-tale signs of bruises. His eyes snapped back to her face. The wisp of grey hair was gone.

"What," she said, a bit of panic rising in her voice. She noticed that Clay was looking her up and down with an expression of disbelief on his face. "I'm sorry, Clay. I . . . I—"

"Eva, I'm sorry. It isn't you. It's this god-damned world we live in." He looked down and then back to Eva. "What's going on? I don't know what's real anymore." He let out a deep sigh. "I don't know who Lillian is anymore, if she's even real. I'm not even sure I know who I am anymore, or what just happened, except— " he paused. "Except that it was about the most real thing that I've done since . . . I can't even remember when. And right now, I can't figure out if that's' a good thing or a bad thing." He finished zipping his pants and left the bedroom and poured himself a drink from her freezer. Eva followed him. She was no longer crying. She wiped the moisture from her eyes.

"Better pour two," she said as she curled down onto the couch. She tucked her feet close, flipping the end of the robe to cover them. Clay handed her a full glass of vodka and sat down on a chair across from her. He started to speak but she stopped him. "I'm not inoculated, Clay. It's a long story, but you may as well know the truth."

"But, I don't understand. Your hair, your bruises . . . ." He looked around again, unsure of where he was or if he was, perhaps, dreaming.

"Of course, very few people know, or would even need to, but they've developed ways for people who aren't inoculated to stay relatively young looking."

Clay's eyes opened wider. "People who aren't inoculated?"

Eva raised her hand and shook her head. "We'll get there, but yes, Clay. There are, I think, a few of us who are not inoculated. I have no idea how many. Maybe it's just me, but I don't think so."

"But if you're not inoculated," Clay started.

"It's called ReGenbots, kind of a topical PreVentall" she interrupted. "I just applied some to my face and hair, which explains why I don't have the bags under my eyes or the grey hair any longer. Or the bruises." Her hand reached to her chest and she moved it slowly over her breast. "It's not permanent. I have to continually administer doses to myself to give the appearance that I'm inoculated."

Eva took a long drink from her glass. "Sometimes things don't quite work out like planned. Like tonight, for instance, when you came here I had just arrived from Israel. It's been a crazy week, and I haven't been able apply any ReGenbot since . . . since." She took another drink. "Anyway, that's why my hair and face looked like they did." She laughed under her breath. "It's no wonder I received a couple of strange looks at the airport." She turned to look out the widow. "Christ! If word gets around, they'll be pretty damn mad."

"Who are 'they'?"

"I'm sorry, Clay. It's a long and complicated matter. Let me start again. You know I've been doing some research for DARPA."

Clay nodded.

"You also know that I don't have a Replicator or an iMeme or any of the equipment the government deems to be 'too dangerous' for me to possess," Eva said, making quotes in the air with her fingers. "Well, because of what I do, I have access to and knowledge of incredible amounts of very important information. I can, for example, provide someone with the technology required to make a Replicator. As you may know, the government does not want Replicator technology to leave this country. They somehow feel that it would be bad for society – or at least for America – for the entire planet to possess Replicators. I have my own ideas on why, but since I, and I can only presume a few dozen other folks, have the ability to pass certain information out of the country, the government has decided that we should not be inoculated. That way, we cannot be duplicated and carry on with our duties here in the U.S. while a Replicon of ourselves is off in another part of the world committing acts of treason. Of course, that was their story. They sold us on the need to protect ourselves from possible harm, so that we weren't kidnapped, duplicated without our consent and used for ill-gain. So while the rest of the world is out there being young, I am not. I may look it, but I'm not inoculated. I can, whether I so choose or not, grow old, or ill, or . . . or who knows what else.'

Clay sat back in his chair. He was dumbfounded.

"No one is supposed to know this, of course, but I trust you, Clay. You've always been there for me and you've taught me so much over the years." Eva sat up on the couch, pulling the robe tightly around her. She looked away from Clay briefly before returning her attention back to him.

"And as long as I'm making confessions here," she paused. "I should tell you that I'm not sorry for what happened just now. Between you and me." Although Eva had hesitated at stating her feelings out loud, now that they were out, she felt relieved and, somehow, relaxed. "And I'm sorry that I dragged you into my crazy fantasy world. You know that I love Lillian and would never do anything to hurt her, or you. I certainly never expected to tell you my feelings, let alone . . . actually . . . ." She looked toward the bedroom. "Although I must admit that it was quite incredible." She blushed slightly. "I'm sorry. That was totally uncalled for. I apologize."

Clay stood up. "No, you don't need to apologize. It was me who let you kiss me. Twice. And I didn't end it there. I let it continue. It was me who took you down that road. I'm still not sure what came over me." Clay paused. "Don't get me wrong. What happened here was real and somehow . . . I'm not sure how to say this, but . . . it was somehow familiar, like we've done this before." He shook his head. "I know, that's ridiculous, I'm sorry. I . . . I don't want to hurt you, Eva. There's a lot going on inside my head right now, but I don't feel bad about what happened, what we did here." Clay took a long drink, finishing his glass and rose. He reached to take Eva's half-full glass.

"Wait a moment," she said, placing her hand on his.

He was conflicted inside. He loved Lillian, but he was not in love with her any longer. He couldn't place his finger on when it happened, but sometime after she was inoculated, their whole relationship changed. They each went in their own directions and never really stopped to look back. They played the role of husband and wife, dinners, family events, sex. But somewhere along the way it was no longer real. He looked Eva over, sitting on the couch before him. He was suddenly filled with a passion so strong he had to turn away. His head felt like it was going to explode.

"Clay, I want you to know that nothing that happened here will hurt Lillian or you. I can't tell you how I know this, but you'll have to trust me. Everything will be all right. I promise."

Clay looked at Eva with some doubt. *How could what just happened not change things?* he thought. He struggled once again with the rising passion he felt towards Eva and the guilt lurking beneath for cheating on his wife. He closed his eyes to think about these things, to try to sort things out a little but Eva began talking again.

"What the government is doing here is evil, Clay. There are things going on that I am a part of, things that are bad. And I'm going to change it." Clay looked down at her, trying to comprehend what she was saying. "Do you remember the night we met downtown, when you were invited to meet with Eli?" She immediately recognized that Clay didn't know what she was talking about. "I'm sorry. The night you were going to meet . . . what did you call him . . . The Walker or-"

"You mean the Walking Man?" Clay sat down on the couch beside Eva. "Yes, I remember. Why?"

"His name is Eli Nakosh. He was a scientist with MIT before he quit and dropped out of society. At first, most of his colleagues thought that he had snapped, that he went a little crazy. But that's not true. He was involved in some of the early and important work in nanotechnology. He was contacted by the government and they sat him down and drew him a picture of the future. It was fantastic: good health, abundant food, free energy – all the things we have come to know now. But he wasn't satisfied with what was offered. For every success, he knew there must be a failure. And for every benefit to man, there would be an equal and opposite harm. He worked with DARPA for a while, helping make some of the fantastic advances that allowed the present to be possible. But he kept a vigilant eye out for the downside. Science must always confront the ethics of dual-purpose research: discoveries that can both benefit mankind and harm it. Think of nuclear fission - a source of green energy, but also untold destruction. At first, the tradeoffs were easy. Two steps forward oftentimes led to only one step back.

"One night he was working late and Colonel Leeds came to the lab. Eli had only met him once prior, but he had already formed a distrust of the Colonel. Professor Ho was also present. Leeds and Ho had a serious conversation outlining DARPA's whole plan."

"And they just threw this out there? With the Walking Man in the lab? I don't know if I— "

"Dr. Nakosh said they were speaking in Mandarin. Apparently neither Ho nor Leeds knew that Nakosh was fluent in the language," Eva interrupted.

"Oh," Clay paused. "So what are they planning?" Clay asked.

"First the Replicators. Then PreventAll. Society would all support the activities. Good health, abundant food supply, free energy. And while these benefits spread out to Europe and some of Asia, there were vast areas that would continue to be denied the benefits. And not because of existing infrastructure issues, although that certainly made some places unwitting beneficiaries of adverse treatment, but the intentional exclusion of certain areas for the sole purpose of control. By keeping a certain level of fear against some evil, society would continue to rely upon government to protect them. And by doing so, those in power would ensure that they would remain in power."

"But why would anyone go to the trouble of trying to stay in power?" Clay asked. "If everyone has what they need, then who cares? They aren't making themselves any better off than anyone else - everyone has more than they could ever want. I don't understand."

"It's about power. And . . . well, if you know anything about Leeds, you'd know he's a cruel and sadistic person. I think that even with everything, they still wanted power. There's more to it than that, but Professor Nakosh wouldn't comment further. Nakosh started to fudge his work a little. He wasn't able to deliver on promises and DARPA eventually let him go. Leeds surmised that perhaps Nakosh had just been lucky in his earlier work, that he had not been as bright as everyone had thought, but Leeds was wrong. Nakosh was, and still is, brilliant. He worked out a much larger plan. One that he's still executing. And those in control aren't as smart as they think they are. Those of us who have not been inoculated may not be able to spread our knowledge as easily as we could replicate ourselves, but we have other advantages. They can stop us from using our technology to spread our knowledge, but by not being inoculated its much more difficult for them to control us, they cannot track our every move. They cannot know our thoughts and our feelings. Yes, Clay. If the government wants to know what you are thinking, they can find out. Every thought you have can now be uploaded to the cloud. Right now, it's voluntary – I'm sure you've heard about people doing it. It's BeMe. The government will let people become comfortable with it and once it's used by everyone, they can simply switch it to full on, 24/7. We're not free. We're caged animals. Happy animals, for the most part, but that's because too many people don't ask questions. They don't think too deeply about anything. But why should they? They're happy. Everything seems good. But it's what Professor Nakosh hasn't said that worries me. But Nakosh has a plan."

Clay perked up. "A plan?"

"He hasn't really shared it with me, but we're going to prevent that. Eli, the group and I."

"Why are you telling me this? What are you going to do?"

"I'm telling you this because it doesn't matter." She stood up and took Clay's empty glass as she walked back to the kitchen. She set both glasses in the sink and pulled out a couple of shot glasses from her cabinet. She turned back to Clay, "I've got some amazing limoncello. I think it'll help clear your mind." She could see the questions racing through his mind. She uncorked the bottle and, glancing back and seeing Clay staring out of the window, she reached into the front pocket of her robe and pulled out a small vial. Her eyes welled with tears. She dumped the contents into one of the shot glasses and topped it off with the limoncello. She placed it before Clay, still holding the second shot glass of limoncello.

"Cheers," she said as she lifted her glass and drank down the contents. A tear ran down her cheek.

"Cheers? That's an odd thought," Clay responded. He knocked back the drink and as he did so, he fell backwards off the couch and to the floor. His eyes grew large and he struggled to breathe. "What's . . . I . . ," he labored to speak as his eyes rolled back into his head. He stiffened and then his body went limp.

Eva took the glass that lay on the floor. She wiped the tear from her cheek and walked back to the kitchen. She poured herself another drink and leaned against the counter, staring at Clay's lifeless body. "I'm sorry Clay. You know too much." She washed the glasses and put them away. She straightened out her place erasing any evidence of her recent presence. She eyed the backpack still lying on the floor near her front door. She stuffed in several more items, and quickly dressed. She took one final look around pausing at Clay's lifeless body. She wiped her eyes one last time and picked up her backpack. She exited the unit and thought briefly of what happened. She paused, but did not look back.

The End

Acknowledgements

I would like to acknowledge the several people who have asked about this book and who showed curiosity and devotion to the first book of this series, Leviticus. I'd also like to acknowledge the love and encouragement from my family and friends. Finally, I'd like to acknowledge my yogi, Muriel Quinn, who helped start me on the path to discovery of the Divine in each of us. May the energy of Shakti find her rightful place in our world today and help us heal.

ABOUT THE AUTHOR

Daniel Seltzer holds a J.D. degree and a BA in English. He previously worked at a major university researching the ethical, legal and social implications ("ELSI") of nanotechnology. It was while working there that the idea for this story first took shape.

www.ingramcontent.com/pod-product-compliance
Lightning Source LLC
Chambersburg PA
CBHW061152170626
46809CB00003B/1068

* 9 7 8 0 9 8 9 8 0 4 5 2 3 *